Making Marie Curie

SCIENCE.CULTURE

A series edited by Adrian Johns

OTHER SCIENCE.CULTURE SERIES TITLES AVAILABLE:

The Scientific Revolution by Steven Shapin (1996)

Putting Science in Its Place by David N. Livingstone (2003)

Human-Built World by Thomas P. Hughes (2004)

The Intelligibility of Nature by Peter Dear (2006)

Everyday Technology by David Arnold (2013)

The Gaia Hypothesis by Michael Ruse (2013)

Making Marie Curie

Intellectual Property and Celebrity Culture
in an Age of Information

EVA HEMMUNGS WIRTÉN

The University of Chicago Press ❋ Chicago and London

EVA HEMMUNGS
WIRTÉN is professor
of mediated culture at
Linköping University,
Sweden. She is the
author of *Terms of Use:
Negotiating the Jungle of the
Intellectual Commons* and
*No Trespassing: Authorship,
Intellectual Property
Rights, and the Boundaries
of Globalization.*

The University of Chicago Press, Chicago 60637
The University of Chicago Press, Ltd., London
© 2015 by Eva Hemmungs Wirtén
All rights reserved. Published 2015.
Printed in the United States of America

24 23 22 21 20 19 18 17 16 15   1 2 3 4 5

ISBN-13: 978-0-226-23584-4 (cloth)
ISBN-13: 978-0-226-23598-1 (e-book)
DOI: 10.7208/chicago/9780226235981.001.0001

Library of Congress Cataloging-in-Publication Data
Hemmungs Wirtén, Eva, author.
Making Marie Curie : intellectual property and celebrity
culture in an age of information / Eva Hemmungs
Wirtén.
pages ; cm. — (Science.culture)
Includes bibliographical references and index.
ISBN 978-0-226-23584-4 (cloth : alk. paper) —
ISBN 0-226-23584-X (cloth : alk. paper) —
ISBN 978-0-226-23598-1 (e-book)   1. Curie, Marie,
1867–1934.   2. Women chemists—France.   3. Celebrities—
France.   4. Women in science.   5. Intellectual property—
History.   I. Title.   II. Series: Science.culture.
QD22.C8H46 2015
540.92—dc23
[B]   2014026044

♾ This paper meets the requirements of ANSI/NISO
z39.48-1992 (Permanence of Paper).

To Per

Contents

Introduction

In 2009, L'Oréal commissioned a poll from the *New Scientist* to crown the "Most Inspirational Female Scientist of All Time." An unlikely survey for a cosmetic giant, perhaps, if not for the fact that it was sponsored by the L'Oréal UNESCO "For Women in Science" program.[1] The result was something of a foregone conclusion. Marie Curie (1867–1934) demolished the competition, beating her runner-up Rosalind Franklin with double the number of votes. By virtue of having discovered and later isolated radium, and then coining the term *radioactivity* for the new science she and her husband Pierre introduced to the world, Curie remains—along with Albert Einstein—the most instantly recognizable face of modern science. So far the only woman twice awarded the Nobel Prize, her 1903 and 1911 distinctions assure her a permanent seat on the Mount Olympus of science.

Children read about Marie Curie's accomplishments in school, learning just how far perseverance and commitment can take you. For each new generation, the same lesson applies: the sky is the limit, even for girls. Countless adolescents who dream about Life in the Laboratory consider the Polish-French scientist a role model. Curie's private and professional life continues to fascinate and supply steady demand for new biographies. No textbook, dictionary, or exhaustive encyclopedia of twentieth-century science would be considered complete without her in it. Indeed, her fame is such that in 2005, the television audience of France 2 ranked her fourth on the list of the ten "Greatest Frenchmen of All Time," trailing De Gaulle, Pasteur, and l'Abbé Pierre, but preceding comedian Coluche.[2] Only a few years previously Curie was crowned the most admired historical French person, well ahead of World War II resistance hero Jean Moulin and *sainte patronne* Jeanne d'Arc.[3]

Admired, revered, idolized. Character-wise, there is no end in sight to the praise. But what about her contribution to science? Here, the verdict is less honorific. Laurent Lemire, one of Curie's French biographers, suggests that she might be the victim of an "Anglo-Saxon depreciation logic" fueled by cultural incomprehension and competition.[4] According to Robert Merton, there never was a Curiean era in the way that there was a Newtonian, Darwinian, Freudian, Einsteinian, or Keynesian one.[5] Be that as it may. If we apply Merton's potential to become an eponym—a name powerful enough to label an entire epoch's worth of knowledge—as a standard, then we will be limited to a very small sample of people. And they would all be men.

In contrast to Newton, Darwin, Freud, Einstein, Keynes, or any other Great Male Scientist targeted for a storyline using The Man to get at The World, Curie resists abstraction. The male scientist, for all his idiosyncrasies, retains his ability to function as a catalyst for *generalizable* observations about science. He *fits*, even as a misfit. Curie, on the other hand, circulates in the closed loop reserved for a specific historical actor, whose experience as a woman is so extraordinary that it cannot be abstracted or generalized. As a result, the only story Marie Curie can tell us is the one about her. Period. Compared to a sad Mary Poppins, labeled the Edith Piaf of radioactivity, even likened to Victor Hugo's Cosette, Curie may be malleable enough to be anyone, real or fictitious, as long as she stays unique, one-of-a-kind, beyond comparison.[6] Perhaps this is why, eighty years after her death, Curie still comes to us as a remarkably one-dimensional person, caught between the two twilight zones of the "dismissive and the hagiographic."[7] As much as possible, I have tried to resist any diminishing logic that situates her as a woman first, a person only second, and a catalyst for generalizable observations on the conditions of modern science a distant third.

We make room for Marie Curie in our collective consciousness for all sorts of reasons. My own motives never included trying to unearth the "real Curie" behind a century's worth of representational bounty. Authenticity searches are hopelessly quixotic to begin with, but even more to the point, I have never considered the representational bounty surrounding Curie a curtain hiding something *really* interesting. As the title suggests, my concern is precisely with the work that has gone into (and that continues to go into) the making of Marie Curie. This means recognizing that the hybrid traces of laboratory notebooks, articles, papers, patents, legal doctrine, advertising, penny press clippings, and popular science articles as well as the polls, encyclopedias, top-ten lists, biographies, and biopics are as relevant to the ongoing cultural construction of Curie as any biographical (or even scientific) so-called fact.

>>><<<

The inspiration for this book is straightforward enough. It can be traced back to my longstanding interest in the contentious role of intellectual property in late modern society. An interest historically motivated, not because I expect the past to serve up neat answers to the predicaments of the present, but because I share Steven Shapin's belief that we write about the past as an expression of present concerns and that "we *can* write about the past to find out about how it came to be that we live as we now do, and, indeed, for giving better descriptions of the way we live now."[8]

But exactly what does a scientific life look like today? And how can someone like Marie Curie possibly tell us something new about it? My answer will come in the shape of the three combinable motifs that structure the narrative of this book: the impact of intellectual property in the domain of science and research; the emergence of celebrity culture and its role in shaping the image of the scientist; and finally, the question of how to organize scientific information as part of the modern infrastructure of knowledge. Although other choices could have been made, I take these three concerns to be central to contemporary scientific life. So they were to Marie Curie's. Simply put, it is my hope that the juxtaposition of these three entities does end up giving a better description of "the way we live now."

Intellectual property has become ubiquitous throughout the academy, despite a long history of being considered ideologically antithetical to traditional academic values of openness and sharing. Patents focus on innovation, trademarks on information, and copyright primarily on cultural expressions: together they constitute the benchmark regulation regime for universities and higher education. But this is only the benchmark *formal* regulation regime, because as in all social relations, the law does not reach everywhere. *Informal* codes of conduct and norms continue to have a presence in university culture and beyond. Nonetheless, lengthy negotiations regarding the scope of the Big Three—alone or in combination—await any scientist embarking on a new project, whether slotted in the "pure" or "applied" category. From publish-or-perish to patent-or-perish, a jungle of licenses, trade secrets, and confidentiality agreements has increasingly turned laboratories into walled and privatized spaces within universities.

More than a century ago, the Curies took a different view and chose *not* to patent radium. Refreshingly out of sync with current intellectual property reflexes, but hardly controversial at the time, this action of nonaction remains a watershed moment in the historical timeline of modern science. An almost mythological aura surrounds their disavowal of patenting and simultaneous embrace of "science for science's sake." A stance with great longevity, it resonates well with the ideologies of the digital openness movement. My first chapter, "Me, Myself, I: In the Interest of Disinterestedness," opens with Marie Curie's own account of the decision. From there, the story continues with an exploration of Curie's careful calibration of interest and disinterestedness; it was a precarious balancing act between different value systems that produced the Curies' famous science ethos. When I realized that their relationship with intellectual property—especially her relationship—was much more complex than I had suspected, it opened the door into a largely forgotten history. As a married woman during the period when the Curies discovered and isolated radium, Marie Curie was not a person. To flesh out the potential intellectual property strategies of someone not able to hold property, we must broaden our understanding of how autonomy and authority follow from the "sexing" mechanisms of the law.

The construction of personhood is a lead-in as good as any to the second aspect of contemporary science I have tried to address: celebrity culture. At first blush, science and celebrity culture appear worlds apart. Not so. The second chapter introduces a competing arena to the academic, one where scientific authority and credibility also play important roles, albeit with a different rhetoric. During Curie's time, the mass press and journals popularizing science for a general readership provided a new platform from which the full-blown celebrity scientist materialized. Today, academic superstars are made through a cornucopia of technologies and social media, where the university and the scientist both have become brands that stand and fall together. To consider the scientist's name not only as a guarantee of truth but as a trademarked commodity takes us into the domain of fame and celebrity. An endless line of interchangeable television reality shows such as *Big Brother* or *The Bachelorette* has put a new spin on the dictum "famous for being famous." As a rule, we rarely find scientists in such a category, traditionally associated with the ups and downs of Hollywood actors, sport stars, and politicians. Curie is one exception, and Albert Einstein, a list regular on the *Forbes* annual tally of top-earning dead celebrities, another.[9] Far behind the top-grossing Michael Jackson, in 2013 Einstein still pulled in a respectable $10 million to the Hebrew University of Jerusalem, the owner of his intellectual property.

While Curie is not on the *Forbes* list and stands in Einstein's shadow when it comes to sheer merchandizing clout, she was a celebrity practically from the day she and Pierre Curie discovered radium and definitely from the time they received the Nobel Prize in 1903. If we want to understand how Curie became a scientific *persona*—the culturally produced trace or copy of the person—we need to look no further than to this intersection of property, legal person, and celebrity culture. In Curie's case, this triumvirate would never form a more explosive combination than during the year 1911. Chapter 2, "Scandal, Slander, and Science: Surviving 1911," looks at two related events that pushed her into the spotlight during her second Nobel year. Her *annus horribilis* opened with her unsuccessful candidacy for the Académie des sciences and drew to a close with five duels fought over her. When news of the romance between Curie, widowed since Pierre Curie's accidental death in 1906, and the married Paul Langevin detonated in public, insults and verbal abuse eventually led nine journalists and one scientist (Langevin himself) to polish their blades and pistols to defend tradition, honor, and Frenchness. Much was at stake for the nationalist, right-wing press leading the assault on Curie: a whole way of life, in fact.

Curie the person could not have become Curie the persona without an audience. As thirsty for scandal as the public was in 1911, they became just as big-hearted a decade later. And no audience was as appreciative of Curie as the American. When Curie made her first trip to the United States in 1921, it was to receive a gift of one gram of radium. The price tag? One hundred thousand dollars, or a sum in excess of a million dollars in 2014. Thanks to a brilliant campaign by Missy Brown Meloney, editor of the *Delineator* and unwavering supporter of Curie, U.S. women had collected enough money to buy Curie a gram of the world's most expensive material. Was it an unbroken chain that connected the Curies' giving away of radium two decades earlier to the reciprocal giving of hundreds, thousands of U.S. women? As we see from the third chapter, "The Gift(s) That Kept on Giving: Circulating Radium and Curie," gifting was a bit more complicated than it sounded. Curie's relationship with Missy Brown Meloney gives us a glimpse not only into their affinity for one another, but also into the impressive networks and alliances the two women drew on as a team. In the end, Curie and Meloney's friendship reveals the extent to which our prejudices prevent us from recognizing that science is made in the most unexpected of places. Meloney's impressive networks—personal and professional—had an undeniable impact on Curie's consolidation as a scientist.

So did another development during the 1920s. The need to organize what was perceived as an abundance of scientific information preoccupied Curie during the last decade of her life. It is also the book's third and final motif. The vertigo that presents itself when we are faced with the choice of databases, indexes, and bibliographies that point to the books, articles, proceedings, and patents that constitute the outputs of science sounds like a very familiar late modern challenge. How do scientists find what they need to read? How are they themselves found by the same systems? And how does all of this translate into research funding? There was a similar feeling of information overload during the interwar years. In the fourth chapter, "Intellectuals of the World, Unite! Curie and the League of Nations," we will meet Curie as a member of the Commission internationale de coopération intellectuelle (CICI). By looking at two of the topics CICI prioritized—bibliography and scientific property—we will see the contours of the modern information order we know today begin to emerge. It is only logical, given her longstanding commitment to "pure" science and disinterestedness through publishing, that Curie would be highly invested in CICI's work on bibliography and international information sharing. This interest coincided with those of the two Belgian visionaries Paul Otlet and Henri La Fontaine, who more than anyone during this time pursued the overlaps between the bibliographic and science communities. That Curie would actively promote scientific property, that highly controversial CICI proposal that never saw the light of day, is more unexpected.

If a microhistorical hook—a decision, five duels, a trip to the United States, a committee—inspired each of the four main chapters, the ambition of the epilogue is to synthesize and draw out some of the more pertinent if idiosyncratic links between the past and the present. But before arriving at that point, I first trace a career spanning two centuries and one world war, look at the arrival of celebrity culture, and consider the work that went into the international management of scientific information. That journey will I hope provide a composite picture not only of the making of Marie Curie but, in however small a measure, also of the making of modern science.

1

Me, Myself, I: In the Interest of Disinterestedness

"In science, we must be interested in things, not in persons."[1] It is precisely the kind of statement you would expect from Marie Curie, one Eve Curie claimed her mother used as a generic retort when dealing with reporters too inquisitive for their own good. Upbraiding journalists for losing sight of what she felt constituted a more appropriate target for their interest, Marie Curie knew better than most that science was nothing if not about persons. So when she was approached in 1920 to write a biography of her late husband Pierre for the book series "Les Grands Hommes de France,"[2] years of firsthand experience as an international celebrity had taught her two things. First, she was in the position to ensure Pierre Curie a well-deserved place in the company of immortals such as Descartes, Talleyrand, and Racine. Second, there were real advantages to that quintessential text "about-persons-rather-than-things" that science on the whole did better without. Perhaps reminding the world of her husband's accomplishments was motivation enough to sign the publisher's contract. And yet, because his work and life were so tightly bound to hers, she was also offered—under the most acceptable of forms—the possibility of overseeing her own legacy.

Managing their public personas in print was something scientists were increasingly willing to do, and readers were eager to take it all in. Young persons especially needed to appreciate what it meant "to be devoted to science," as Henry Neumann framed his request for Curie's permission to include an excerpt from the U.S. edition of *Pierre Curie* in his book of readings due from the *Atlantic Monthly*.[3]

The biography outlined her spouse's personal qualities and followed his early achievements in painstaking detail. And while its author modestly questioned her ability to accurately depict her husband's childhood, few would have entertained the idea that this particular book could have been penned by anyone else. So when the narrative provided her with an opportunity to make one of the few openly programmatic statements about how the famous husband-and-wife team viewed their work, she made the most of it.

> Our investigations had started a general scientific movement, and similar work was being undertaken in other countries. Toward these efforts Pierre Curie maintained a most disinterested and liberal attitude. With my agreement he refused to draw any material profit from our discovery. We took no copyright, and published without reserve all the results of our research, as well as the exact processes of the preparation of radium. In addition, we gave to those interested whatever information they asked of us. This was of great benefit to the radium industry, which could thus develop in full freedom, first in France, then in foreign countries, and furnish to scientists and to physicians the products which they needed. This industry still employs to-day, with scarcely any modifications, the processes indicated by us.[4]

As she enumerated the dos and the don'ts, Curie situated scientific practice within a gift/market dichotomy with two distinct systems of credit and reward. Intellectual property represented an "interested" perspective where you "reserve advantage." Choosing to "publish without reserve" and keeping "no detail secret" instead epitomized the values of disinterestedness. Abstaining from pro-

FIGURE 1. Caricature of Pierre and Marie Curie appearing in *Vanity Fair*, December, 22, 1904. With permission from the Musée Curie (Association Curie et Joliot-Curie), Paris.

prietary shackles on radium spurred more innovative activity in both science and industry, and not less. The industry could then develop "in full freedom, first in France, then in foreign countries." This is as close as we get to a will and testament about the discovery of radium and the science of radioactivity, *the* leitmotif Curie wanted associated with her scientific legacy. Material profit was refused, but on the other hand publishing took place without reserve. No advantage was reserved in industrial application, but no detail was kept secret and information given freely. Diligent readers might find the expression "took no copyright" slightly odd given the circumstances. It could have been a blunder on the part of the American translators, Charlotte and Vernon Kellogg, because even if copyright had to be registered in the United States at the time, the original French text does refer to the more logical *brevets* (patents).[5]

And what was the result of their action of nonaction? Not the opening up of science's new great unknown, but the blossoming of a radium *industry*. And all of this because the Curies subscribed to the value of *disinterestedness*. Scientific discovery does not come about for any ulterior motive than curiosity; it serves no other master than itself and is its own reward; disinterestedness goes together with universalism, communism, and organized skepticism, Robert Merton's classical imperatives of modern science. But look beyond the altruistic surface and you find a remarkably strong claim to radium. Everything—in the science *and* industry of radium—began and ended with the act of relinquishment, an act that paradoxically ensured the Curies the strongest possible entitlement to the fluorescent new element they had given away.

Taking this quotation as point of departure, this chapter aims to understand something of how this powerful representational dichotomy between patenting and publishing, between the "pure" and the "applied" ended up the way it did in *Pierre Curie*. I hope to show that Marie Curie's rhetorical choices resulted from a set of claiming strategies that secured control over radium *outside* formal intellectual property regimes. Both the patent and the article, for instance, the two texts Curie situated on each side of the pure/applied dichotomy, are claim-making texts. Rather than considering how they diverge, I want to think about the space where they converge. Choices were made in respect to the authorship, ownership, and control of radium that hold an important key to a more nuanced understanding of how the Curie persona would evolve. But in order to unpack just how much interest went into scientific disinterestedness, we need to get a few biographical data out of the way first.

I

Pierre Curie and Marya Skłodowska married on July 26, 1895. With the exception of two bicycles given to them as a gift, neither husband nor wife had any possessions to speak of and no marriage contract was signed. Four years previously, the future Madame Curie had arrived in Paris to study science at the Sorbonne, where she graduated top of her class for the *licence ès sciences physiques* in 1893 and second the following year for the *licence ès mathématiques*. Enrolled at a time when the French university system was in the nano-embryonic stages of gender equality, Marya Skłodowska was not merely one of very few females in an overwhelmingly male student population; she was a Polish woman among French men. Initially, such status may have provided an unexpected bonus as far as access to the university was concerned, but as we shall see in the next chapter, later it proved a lethal combination of gender and nationality her enemies claimed precluded her from contributing anything of value to French science or to French society in general.

Fortunately for her, she knew nothing of what lay ahead on that score when she successfully graduated from the Sorbonne in 1894. Thanks to Gabriel Lippman, professor of physics at the same school and friend of both Curies, she received a grant from the Société d'encouragement de l'industrie nationale (SEIN) to study the magnetic properties of steel. During her work for SEIN—since 1801 an important initiator of industry-based research in France—she met her husband-to-be, who defended his thesis at the Sorbonne in the spring of 1895, becoming *docteur ès sciences physiques*. Their first child, Irène, was born on September 12, 1897.

Only a few months after giving birth, Marie Curie embarked on her thesis, the topic of which caused her some initial hesitation. She could have continued her research on steel, but two recent and sensational events turned her attention elsewhere. First, there was Wilhelm Röntgen's 1895 discovery of what he "for the sake of brevity" called X-rays.[6] Then, Henri Becquerel, professor at the École polytechnique, accidentally forgot a pack of uranium salts in a drawer, only to find that they had emitted an unknown, spontaneous radiation on the photographic plate where he had left them. X-rays produced evocative images capturing the imagination of the broader public, but uranium salts were expensive to acquire, their visual appeal far less spectacular and their practical application uncertain. Becquerel published seven papers on his serendipitous discovery in 1896 but only two in 1897,[7] and the scientific community's interest in it seemed to be waning. Not so Marie Curie's. "The study of this phenomenon seemed to us very attractive," she wrote, "and all the more so because the question was entirely new and nothing yet had been written upon it. I decided to undertake an investigation of it."[8]

The formidable twosome of *Pierre Curie* materialized during a six-year period beginning in 1897 and ending with the 1903 Nobel Prize. This is when the collaborative "I" and the "we" discovered (1898) and isolated (1902) radium, work that first took place in a small, glassed-in space used as a storage room for machines, and then in an old, abandoned shed. Both makeshift laboratories were located on the premises of the École municipale de physique et de chimie industrielles (EPCI), Pierre Curie's academic home for more than twenty-three years. When EPCI opened its doors in 1882, it was in response to a somewhat brutal French wakeup call following the 1878 Exposition universelle, where the international contributions had proven unexpectedly competitive. As always, France first measured her strengths against Germany.

And came up short. The blame for lagging behind, especially in chemistry, was placed on a lack of professional schools and adequate training. Since these concerns were exacerbated by the loss of Alsace to the archenemy, there was some logic to EPCI's being the brainchild of Charles Lauth, an Alsatian chemist. Pierre Curie, about to make a name for himself in the scientific community, was perfectly matched with this new institution outside the traditional *grandes écoles*. Home-schooled, never attending the elite École normale supérieure or École polytechnique, he was described by his biographers as a bit of a loner and outsider, an image he to some extent appears to have cultivated himself.

The laboratory housed gentlemanly pursuits of disinterested ambition, while simultaneously providing industry with high hopes for a less "pure" and more "applied" future that would benefit the nation in the competitive, international race for industrial supremacy. But the laboratory was also a legal space, one where bodies, laws, and texts clashed and overlapped. As an illustration of how the discovery of radium and the Curies' claims to it had something to do with the law, I want to use the 1804 Code Civil and the 1844 Loi sur les brevets d'invention as a backdrop for this chapter. The Code Civil had set up a detailed, famously patriarchal system that governed the perimeters and rationales, conduct and misconducts relating to family life. The 1844 Loi sur les brevets d'invention captured a different set of relations, one that placed itself in the sphere of innovation and progress, a public and professional sphere separate from that of family life and domestic nitty-gritty. Rather than uphold any artificial distinction between the two spheres of private/public, I prefer to situate the laboratory as a permeable continuum between "pure" and "applied" science, between interest and disinterestedness, between gift and market, but also between the Code Civil's "petit monde" and the "grand monde" suggested by the Loi sur les brevets d'invention.

In the Code Civil enforced at the time of the Curies' life together, article 213 stipulated that "the husband owes protection to his wife, the wife obedience to her husband," an article that was not removed until 1938. Article 1124 of the Code Civil judged Marie Curie *incapable*, a status married women shared with children and the insane. Eugène Pouillet, author of *Traité théorique et pratique des brevets d'invention et de la contrefaçon* (1899), a 1000-page standard treatise on patents reprinted several times, explained: "by *incapables* the law refers to those whose social status, weakness of mind, or presumption of such weakness in effect prevents them from managing their own affairs, and as a consequence, from signing contracts."[9] The clause remained active until 1938. Not until July 13, 1907, did the law recognize married women's right to "the products of their work" and allow them membership in a *syndicat* and the right to receive a pension. The right to vote, however, was not secured until 1944, ten years too late for Marie Curie to be able to exercise her civic duty.

Nevertheless, we still know something about her feelings on French women's suffrage. On July 7, 1932, when the question went before the Senate—ultimately defeated with an overwhelming 253 votes against 40—Curie's name came up during the debate. There was not much the opposing camps agreed on, except perhaps that only three living women belonged to the French elite: Mme Curie, Mme de Noailles, and Mme Colette. Louis Barthou, who admitted never having broached the subject of the vote with Curie, felt sure that if he had, she would *not* have pronounced herself favorably on the topic. From several conversations with the other two women, Barthou concluded they were both "hostile to the 'suffrage féminin.'"[10] *Très bien!* Shouts of acclamation and applause echoed in the Senate. It went without saying that if only three women in France had attained sufficient prominence and sophistication to be trusted with the vote, then it was political suicide to make a blanket extension of the right to all women. In 1932, Curie was a world-famous scientist who rejected all requests that did not meet her benchmark standard of being directly related to "pure science." Here, however, in one of the most controversial of political issues in French society, she did have an opinion. The day after the debate, she wrote to Jules Jeanneney, the president of the Senate, to set the record straight.

To say she was unfavorable to women's suffrage was "surely a misunderstanding," and without pronouncing on the modalities of granting political rights to women, she continued, "I think that the principle is essentially just and that it should be recognized."[11] Jeanneney replied that it was impossible for him to notify the Senate of a correction from someone, "however prominent," who did not belong to the Assembly. But, he added with some insight, "you have, I think, obtained the result you wanted in communicating your letter to the press."[12]

More than thirty years earlier, her marriage to Pierre Curie had subsumed all her property rights under her husband's. Marie Curie was on the path of becoming the first female *docteur* at the Sorbonne, but when it came to holding property she was just like any other married woman during the Third Republic (1870–1940): barred from owning, controlling, and benefiting from either tangible or intangible property. Article 217 of the Code Civil stated: "The wife . . . cannot give, transfer, mortgage, acquire whether free of charge or for a consideration, without the presence of the husband or his consent in writing." All the money Marie Curie received for her work—grants, scholarships and salaries—officially fell under the control of Pierre Curie, who also had all rights over their common property, which he was free to alienate at will. On Irène's birth, article 373 gave him another privilege: "la puissance paternelle," or all rights over any children born in marriage.

The negation of patenting in the *Pierre Curie* narrative becomes a less straightforward proposition when we consider that Marie Curie, wife of Pierre Curie, could not hold property—and I am particularly concerned with the intellectual kind—at the time of their collaboration on radium. This was definitely one aspect of her parents' marriage that Eve Curie never introduced in *Madame Curie*. Perhaps it somehow risked upsetting the carefully constructed balance associated with the most famous husband-wife scientific partnership of all time, a successful collaborative unit seen as the direct result of an equally ideal marriage. Or perhaps Eve Curie wanted us to believe that marital status was an irrelevant factor as far as disinterested science was concerned, an ideal supposedly above bodily restraints.

Most of all, we owe the plotline of a match made in heaven between two enlightened, modern, and equal individuals to the main protagonist of this book. "We lived a very single life," Marie Curie remembered in *Pierre Curie*, "interested in common, as we were, in our laboratory experiments and in the preparation of lectures and examinations."[13] The development of theories, the experiments in the laboratory and the preparation of curricula—not exactly the kinds of topics Madame and Monsieur Martin engaged with over breakfast, but the small talk of this married couple nonetheless. Laboratory work was not only a matter of the lofty meeting of minds, however. Thoughts not only could be transformed into materialities, they *had to be*, because what happened in the laboratory could not stay in the laboratory. You proved priority through papers, lectures, and books, texts that were authored, published, read, and, somewhere along the line, owned.

To say that the law rendered this extraordinary woman extraordinarily ordinary, and to consider property along the lines I suggest, is not to question the Curies' spousal attachment, nor does it belittle their mutual professional respect for one another. Nevertheless, it is important to understand how a public figure, co-discoverer of radium, author of many scientific papers, and Nobel Prize recipient navigated around the fact that she could not "own" any of the intellectual property that led her to those achievements or that resulted from them. It would credit the law with much too much power to argue that it is the great mover and shaker behind every event described in this chapter. Yet by reserving the category of person for men only, the sexing mechanisms of the law fashioned the interpretative possibilities surrounding the authority, autonomy, and authorship that came with their collaboration. Crediting Marie Curie's legal status as a married woman under the Code Civil with importance means contributing to a much-needed expansion of the interpretative horizon within which we consider innovation and the law.

So we have a married couple measuring, scribbling, and weighing their way into print. It was the textual evidence coming out of the "shared interest in all sorts of things"—transferred into various forums and translated into various formats—that made both scientists famous, recognized, and published. But during the period Marie Curie became famous, recognized, and published, she had few legal entitlements and was unable to sign a contract or exercise control over her own grants. During the period when Marie Curie became the public figure we think we know so well, at the time her persona begin to take shape, her formal presence in the public sphere was that of a shadow. In *Droits et devoirs de la femme devant la loi française*, a law treatise from 1884, N.-M. Le Senne laid out the rights and obligations of French women. And as he so succinctly put it, *citoyens français* was a term referring only to men, because it was a category to which women were not admitted.[14] In case we need reminding, the nexus of science, innovation, and intellectual property is historically contingent, not historically given. It would be naïve of us not to expect Curie's strategies as a scientist to develop accordingly.

II

Marie Curie made her first entry in what we today know as the "Discovery Notebooks," on December 16, 1897, when she recorded a test of an apparatus, an electrometer. Historians of science sometimes hold notebooks in almost fetishistic awe. Gerald L. Geison's enthusiasm over Pasteur's jottings is the kind of transcendent *frisson* caused by touching coffee-stained leather covers and frail, ink-blotched paper: "Words cannot fully convey the sense of excitement that comes from turning the pages."[15] The origin, the source, the *urtext*; the recent digitization of the "Discovery Notebooks" is consistent with their mythological status in the timeline of radium and the ongoing digital heritage making of the Curie archives.[16] A more prosaic reason for making them easily accessible on the Web is that they are still radioactive enough to register on the Geiger meter.

Sitting at the workbench and getting everything ready, Marie Curie knew that the finely calibrated instrument she had at her disposal would secure the exact measurements that were the sine qua non of exact science. She had absolute faith in her husband's superior craftsmanship. Like many of his contemporaries, Pierre Curie combined the traditional skills of the *artisan* with the scientific knowledge of the *savant*. Alone and with his brother Jacques, he had patented a number of scientific instruments long before there ever was a "we" on the horizon. Nor was he a passive bystander in their commercialization. Compared with his annual EPCI salary of 6,000 francs, the 1,500 francs his scales and instruments rendered him in annual revenues provided the family with a nice bonus. Even taking into account the couple's grants and various other additional incomes, around the turn of the century licenses still provided a supplement equivalent to a quarter of his primary salary.[17] Patent and instrument, the text and the three-dimensional object, both precede the notebook, announcing and enabling future discoveries and results.

The idea that technical achievement and innovation should be awarded with some sort of property rights originated in fifteenth-century Italian city-states and foremost in Venice, the first city to regulate monopoly patents in 1474. From there, patents evolved over five centuries from royal privilege to regulation by law. On October 6, 1888, when Pierre Curie received *brevet d'invention* 192,377 for a *balance apériodique*, he paid a sum of 1,500 francs to receive protection for fifteen years, the longest period afforded by the 1844 *loi*. In a nice twist of irony, considering his image as an intellectual property refusenik, the Institut national de la propriété industrielle (INPI) today sells a beautiful facsimile print of Pierre Curie's patent as an example of Important French Innovations. But what authority did it confer? Well, not much. Prominently displayed on the *brevet* was the famous caveat "Without governmental guarantee," which stipulated that the patent had been issued to him "without prior examination, at his own risks and perils, without guarantee of the reality, novelty, or merit of the invention or of the fidelity or exactitude of the description."

At least the *brevet* was registered, a prerequisite for any kind of enforcement. Already the Venetian Senate had ruled that when perfected, inventions should be registered, securing the inventor sole benefit for ten years. Infringement was penalized with a 100 ducat fine, and the government reserved the right to appropriate registered innovations. This was long before the introduction of "gatekeeper" functions to determine if a new invention or process truly showed originality and utility. The proviso of the 1836 U.S. Patent Act had reintroduced the requirement of a pre-examination that to this day remains one of the cornerstones of enforcing present standards of novelty, nonobviousness and utility. Whether or not France should fall into line and accept such review before issuing a patent was a hotly contested issue culminating around the time of the 1878 Exposition universelle. Those favorable to a pre-examination held that France was at disadvantage vis-à-vis nations that had such an assessment in place, notably Germany. Those opposed felt that examination exposed the inventor to an arbitrary and uninformed review. It was for the market to determine if an invention was new and useful, and if disputes arose, these should be handled by the courts.

On the left-hand side of Pierre Curie's *brevet* there is another bit of important information, clarifying that the patent was in danger of nullification if exploitation did not occur within two years of the date of issue, or if it ceased for two consecutive years. Innovators were thus under pressure to get their innovations into production and use fairly soon. This could explain why, only a month after being awarded his patent, Pierre Curie signed a contract with the Société central de produits chimiques (SCPC) giving them the exclusive right to the commercial and industrial exploitation of his scale. From patent application to construction to subsequent marketing and sales, SCPC took on all pecuniary responsibility. Paragraph 3 of the contract stated that in publications and exhibitions, the scale would be known as "Balance à lecture directe et à amortissement, Système Curie, Construction de la Société Centrale." Pierre Curie received 10 percent on every sold scale, 20 percent for modifications made by the society on scales and models, and 30 percent on licenses. In the case that the Société decided to cease production of the scale, all rights would revert to Pierre Curie, who, at the very end of the contract, scribbled, "Read and approved!" and then added the date of December 26, 1888.[18]

Aided by her husband's patented instruments, Marie Curie's work progressed well. In the spring of 1898, she was ready to announce her results to a larger scientific constituency, and she turned to the one arena in France authoritative enough to provide her with the amplifying power she needed. Founded in 1647, the Académie des sciences was one of the five academies that together constituted the Institut de France. Known also as the first class among these five, the Académie met every Monday afternoon.

The academy culture was both open and closed at the same time. Closed, because in order to present your findings you had to be a member, and becoming one was the crowning achievement of any scientific career. Once inside, you belonged to an elite, but one whose meetings were covered by regular newspaper reports from the mid-1820s. Partly in an attempt to better control this flow of information into the public sphere, the Académie introduced the weekly *Comptes rendus des séances de l'Académie des sciences* in 1835. Published by Gauthier-Villars, the *Comptes rendus* documented and reported what the Académie had discussed in their previous meeting, as well as printed correspondence and notices of news and expeditions in the tradition of the Royal Society's *Philosophical Transactions*.

By no means the first regular publication of the Académie, the *Comptes rendus* represented a sea change in patterns of scientific communication. Priority had always been of the essence, but in the increasingly transnational arena of trade and communication networks, news of scientific discoveries needed to travel faster and reach readers more efficiently. The transition from book to article was one of speed as well as format. Large-scale, multi-volume treatises that sometimes took years to appear in print were slowly phased out. At forty pages per week, with the maximum length of any member contribution in any issue set at eight and the annual allotment at fifty pages, the *Comptes rendus* still ran an impressive two thousand pages a year, divided into two annual volumes. But the turnaround was the really impressive part; announcements appeared in print only a week following the date of the oral presentation, with the latter still establishing priority.

Since neither Curie was a member of the Académie at this time, somebody who was had to speak for them. According to custom, nonmember communications always appeared last on the meeting agenda. And this is also where we find the Curies' three 1898 *notes* outlining the stages by which they discovered that pitchblende, a byproduct of the radioactive disintegration of uranium, traditionally used for the decoration of Bohemian glassware and viewed as nothing more than waste following this production, contained two new elements, polonium and radium. Since this is a story told many times over, I limit myself to a discussion of how these *notes* established authority over radium by contrasting and combining authorship. Carefully.

The first *note*, written by Marie Curie alone and signed Madame Skłodowska Curie, was read by Gabriel Lippman on April 12, 1898.[19] Nonmembers enlisted members to present their research, not only because it was a requirement, but also because it added credibility. A century away from the present era of hyperauthorship, we can recognize the beginnings of the multiplication of authors we have grown accustomed to in today's science. Gabriel Lippman's name on the *note* is as unerasable as Marie Curie's. And while he did not write the text, he performed it. To the best of our knowledge, his delivery that day did not differ in any radical way from that of previous readings he had made on the behalf of other scientists. There is no evidentiary record of the session, no photographs, no sound recording of his voice. The style, tempo, inflections of the presentation and its impact on the *academiciens* elude us. We remain in the dark about how it felt to be Lippman speaking the "I" of a *female* author/scientist, and how the audience received the enunciation.

In the period between the first and the second *note*, Pierre

Curie abandoned his study of crystals and joined Marie Curie in her work, now concentrating on the elevated radiation displayed by pitchblende. Read by Henri Becquerel on July 18, the second *note* was therefore signed by Pierre and Marie Curie together, the latter now under the name Madame S. Curie. The collaboration with his wife may have been a unique period in his life, but of the thirty papers Pierre Curie published on radioactivity during his career, all but seven were co-authored, primarily with EPCI colleagues like Jacques Danne, André Debierne, and Albert Laborde.[20] This second *note* is more assertive, holding out even more promise for the future. Introducing *radio-active* as a property of the new element they had found in pitchblende was the work of the "we" present in the text, but by using the expression "one of us" they referred back to the findings in Marie Curie's single-authored April *note*.[21] In her 1903 thesis, Curie eschewed the "we" for "I" when taking credit for coining the term radioactive. Aged fifty-seven, she repeated the story in *Pierre Curie*, "I proposed the word radioactivity which has since become generally adopted."[22]

Henri Becquerel, the original source of inspiration for the Curies' research and thus a guarantee of its quality, also read the third and final 1898 *note* on December 26, 1898.[23] It was signed by Pierre Curie, Madame P. Curie, and a third author, Gustave Bémont; the inclusion of Bémont—the "forgotten man of radioactivity"[24]—made it possible for the trio to distance themselves further from the single, female, and therefore less stable author of the first *note*. The instability was not just because of her gender or because she had written the paper alone, but because the paper was a weak link in the claiming chain. Earlier that spring, Gabriel Lippman had read a *note* by a German physicist, G. C. Schmidt, announcing that he—more than a week before Marie Curie—had presented findings similar to those she announced in her first *note*.[25] Thorium was old news by now, but the fact remained; Marie Curie would have to acknowledge Schmidt's priority in her thesis. In doing so, she would also have to admit being second.[26]

The Curies had acknowledged Bémont's help already in the second *note*, but by adding him as co-author they strengthened their findings by authorial accumulation. The third *note* described how they had identified a second element even more radioactive than the polonium they had found and named after Marie Curie's native Poland in July. Similar in appearance to barium, the new element was baptized by them radium, the radioactivity of which, they predicted, "should be enormous." Apart from a self-citation back to the first *note* and Marie Curie's findings—"one of us has shown"—this time the three authors backed up their claims, not retrospectively, but by "a special *note* following our own." In it, Eugène Demarçay confirmed that there was indeed a new element present in what he, in proprietary terms that traveled well from the tangible to the intangible, referred to as "Madame and Monsieur Curie's barium."[27] The scientific claims grew stronger with each *note*, an assertiveness justified not only by the evidentiary traces coming from the laboratory, but also by the different authorial combinations played out in the three *notes*. First Marie Curie is the single author; second, she follows her husband; and in the final *note* she is the second author between two men. Her names were equally flexible: First she is Madame Skłodowska Curie, giving her full Polish name. Next, however, Skłodowska has been abbreviated into a single letter, and she is Madame S. Curie. Finally, following Pierre Curie but preceding Gustave Bémont, she ends up as Madame P. Curie, disappearing behind her husband's name. For each *note*, radium gets stronger and Marie Curie weaker, her authority and authorship increasingly subsumed under Pierre Curie's name. What we have is a veritable training ground for authorship and authority. Whereas Pierre Curie's identity as an author remains stable throughout, Marie Curie's is fluid, tentative, fuzzy in its contours. We see her try out different names, different authorships, depending on the importance of the claim. The more significant the properties of radium to be placed in evidence, the more important it was to assign the discovery to someone who had the authority of personhood. Given these restrictions, it seems only logical that Madame Skłodowska Curie became Madame P. Curie in that third and final *note*.

The *Comptes rendus* consecrated radium in academic terms, but as the daily press provided regular accounts of what had transpired during Académie meetings and snapshot presentations, often under captions such as "A l'Institut," reports of the new element spread beyond the academic community in no time. Radium was a godsend for what in French is known as the *presse de vulgarisation scientifique*. During the year 1889 alone, twenty-five new journals were launched in a genre that sounds oddly and unjustifiably derogatory in English, and interest peaked around 1865–1900. When the Curies received the Royal Society's Davy Medal and the Nobel Prize in 1903, Henri de Parville, *La Nature*'s influential editor-in-chief and "Revue des sciences" columnist in the *Journal des débats politiques et littéraires*, found it ironic that it would take these two awards in order for the general public to discover the discovery of radium. As one who had followed their work during the five preceding years, Parville wanted to sound as initiated as possible, proving just how cutting-edge *La Nature* really was as a leading source of information on science.[28]

X-rays had whetted readers' appetites for more news of spectacular scientific breakthroughs, and from those first experiments at the rue Lhomond barracks, it was clear that radium ticked all the right boxes. Even better for the press, there was a great need for what Lawrence Badash so nicely referred to as "editorial illumination"[29] when it came to radium and radioactivity. As opposed to the rather self-explanatory X-rays, somebody had to make sense of the marvelous properties of this revolutionary French discovery, one that had arrived so opportunely at the very close of one century and the dawning of a new. And Marie Curie was one of those somebodies.

Even before the last and most claim-making of the three *notes* had appeared in the *Comptes rendus*, the editor of the *Revue générale des sciences pures et appliquées*, Louis Olivier, wrote to Marie Curie regarding a piece she had promised him on polonium. When the second *note* came along, he became most insistant that

Curie expand her article to include radium. If she did not, he argued that her contribution would have a "great hole in it," and even went as far as to suggest that he could postpone publication for two weeks to accommodate her alterations.[30] What he received, however, and what later appeared in print, was an article without any radium in it. Ignoring the persistent editor, Curie did a thorough presentation on the Becquerel rays and polonium, and once again mentioned that she had published her first *note* "not yet having learned of M. Schmidt's." Because the publication of the *Revue générale* came so close in time to the December 26 presentation, she was perhaps hesitant to pursue the *vulgarisation* angle more fully when it came to radium. Oliver himself was forced to add that little tidbit as a postscript to her text.[31]

Announcing the discovery of radium to the world meant that the notebooks—temporarily, that is—had served their purpose. Until they reached the *Comptes rendus*, laboratory experiments were considered a private matter. The notebook hypothesis, markings, scribbles, and drawings were essential in key stages of the laboratory work. Records of tedious tests upon tests, notations of minute alternations of measurements, in slightly different combinations or with slightly modified quantities: the notebooks were a superior record of tinkering and testing but hardly blockbuster material. But to survive into cultural heritage they needed the clothes of a different language and different materiality, where they formed relations with other texts, authors, and claims. As both authors and readers of the notebooks, the Curies wrote for themselves, for their assistants, and maybe even for posterity. Notebooks are part of the totality of texts that make up the scientists' textual inheritance, and important in the afterlife of their author(s) not only as a portal into processes of discovery, but also because they are among those texts through which the persona can be reaffirmed.

The *carnets de découverte*, the *notes*, and the *Pierre Curie* biography all form part of the historiography around the Curies' collaboration. The notebooks have a special place in the mythmaking. Perfectly in line with the romanticized narrative of *Madame*

Curie, Eve Curie found it impossible to separate the two authors, the two "amorous beings," whose writing effortlessly "alternates and blends on the pages filled with formula."[32] As opposed to her mother, who went to great lengths to distinguish who did what, Eve Curie would always put greater emphasis on the "we" than the "I." Appearing to provide uncensored insight into the scientist-at-work, it is not easy for the lay reader to decipher the data in the "Discovery Notebooks." But there is help at hand. In an appendix to a 1955 reedition of *Pierre Curie*, Irène Joliot-Curie provided a sort of translation of her parents' notebook notations. Fascinating reading, not so much for making the science more understandable but because we see again how much the making of Curie has depended on the framing of certain key texts in new versions, new forms. Eve concentrated on making the laboratory notebooks part of the Curies' marriage, their love story. As a scientist, Irène took it upon herself to interpret and explain the actions behind her parents' work on radium, stressing that "the object and conclusions of the experiments are indicated by me, but are not found in the text of the notebooks."[33] Eve wrote *Madame Curie*, Irène explained the notebooks, but both did their bit in removing the things so that the persons could be seen more clearly.

III

By the end of 1898, the Curies had a scientific sensation on their hands; radium was signed, sealed, and delivered into print. They had discovered polonium and radium, and coined the term "radio-active." Then what? While they were confident about the two new elements they had found and named, Marie Curie confessed that "to make chemists admit their existence, it was necessary to isolate them."[34] This task, they knew for certain, required the acquisition and treatment of enormous amounts of the mineral. After they had shopped around for—but failed to find—a more suitable laboratory that could house the tons of pitchblende residue they needed for this next step, the EPCI allowed them the use of a derelict outbuilding adjacent to their current laboratory, one that had previously been used by students performing dissections.

The Curies managed to use both patenting and publishing to their advantage. They had published their findings in the all-important *Comptes rendus*, but never turned their backs on an opportunity to build strong alliances with industry. Pierre Curie's instruments—patented and then licensed to organizations such as the SCPC—helped build alliances across the divide of "pure" and "applied" science. Such bridge building was within the purview not only of Pierre Curie. From her SEIN study on steel, which got her her first grant from the Académie, to the establishment of the Fondation Curie in 1920, allowing her to solicit private funds and donations, Marie Curie proved highly adept at developing industry contacts to consolidate her radium legacy. She developed close ties with the burgeoning radium industry later described as the direct beneficiary of the Curies' nonpatenting policy, and later in her career masterfully capitalized on her celebrity status to cement the role of the Radium Institute. As the process of isolation began, the Curies enlisted the support of the SCPC, providing them with chemical products and paying the staff's wages, in exchange for which they obtained a proportion of the extracted radium salts for marketing.

The industry-like conditions at this time made Marie Curie into an industry worker; the one image that has clung to her over the years is that of menial worker rather than inspired scientist, and it stems from the period when she toiled away at the EPCI hangar. Working with twenty kilos of material at a time, Curie later wrote, "it was exhausting work to move the containers about, to transfer the liquids, and to stir for hours at a time, with an iron bar, the boiling material in the cast-iron basin."[35] Following three years of such hard labor in the dilapidated shed on rue Lhomond, on July 21, 1902, Marie Curie announced in the *Comptes rendus* that she had isolated radium and arrived at an atomic weight of 225. But she was not just making an announcement of a scientific breakthrough; she was careful to have her results associated with the preeminent "System Curie" granted protection by the 1888 *brevet*. Her words could have accompanied a sales leaflet promoting the preeminence of her husband's instruments: the *balance apériodique Curie* was "perfectly calibrated, precise to a twentieth of a milligram."[36]

Discovering and isolating radium had presented enormous practical challenges. But the ideological problem now facing the Curies was no less daunting. Paul Lucier describes the double bind of nineteenth-century scientists: "any scientist who patented research put at risk his professional integrity. Still, if a scientist wanted to protect his rights as a discoverer, he would have to patent."[37] But could radium be patented, and if it could, why do it?

Article 30 of the 1844 *loi* declared null and void all patents concerning discoveries or purely scientific conceptions that had no industrial application. But radium promised to be useful in the extreme, and industrial possibilities loomed quickly on the horizon. On the other hand, the first article of the 1844 *loi* stipulated that "all new discovery or invention, in all genres of industry, give to their author, under the conditions and for the period of time given hereafter, the exclusive right to exploit to his profit the same discovery or invention." The invention of new industrial products or the invention of new processes, as well as the improvement of already known methods in order to arrive at an industrial product, all were considered patentable. Patenting would have allowed the Curies to extend their control over radium, but such a step radically negated the openness they and the science community at large were eager to embrace.

Eve Curie fictionalizes the moment when Pierre and Marie Curie have to make a decision about "their" radium as a "five minute's talk" on a Sunday morning. In the American edition of *Madame Curie*, the chapter heading includes the length of the conversation, underlining the importance of her parents' discussion while simultaneously stressing how easy it was for them to arrive at their decision. "We must speak a little about our radium," Eve Curie opens the scene, having her father read a letter from the United States, fold it, and place it in his desk before turning to his wife and telling her that they had two options before them. The first was to describe, without any restrictions, the result of their research including the processes of purification. The second was to consider themselves the proprietaries or inventors of radium and patent the technique of extraction, the very process Marie Curie had developed while Pierre Curie turned to more theoretical work. In Eve Curie's narrative, her father takes great care to present the situation as objectively as possible to her mother, but "it was not his fault, if, in pronouncing words with which he was only slightly familiar, such as 'patent,' 'assure ourselves of the rights,' his voice had a hardly perceptible inflection of scorn." Marie Curie's answer was unequivocal: "It is impossible. It would be contrary to the scientific spirit." Eve Curie lets her father continue listing the advantages patenting would bring: money for them and their children, assured comfort and riches. Marie Curie only briefly considers the possibility of material gain before she again rejects the proposition. Commercial success is of secondary importance to scientific discovery and should be left to others. Radium should be used to alleviate suffering and sickness. "It seems to me impossible to take advantage of that." Not that she needed to convince her husband, Eve Curie writes. Pierre Curie ended the discussion by telling his wife: "I shall write tonight, then, to the American engineers, and give them the information they ask for."[38]

Theirs was hardly a controversial decision, because the Curies simply followed a well-established practice associating patents with industrial application rather than "pure" science. Yet laboratory work was much more involved with patenting and industry than we might infer from Marie Curie's narrative in *Pierre Curie*. Pasteur, for instance, officially as loath to consider commercial exploitation in science as the Curies, took out several patents and benefited economically both from his patent on a bacterial filter and from the international licenses and sales of the anthrax vaccine.[39] Then again, the Curies' colleague and friend Gabriel

Lippman did not patent color photography, allegedly because he felt that state-funded research should not subsidize privatization of an invention.[40]

Of course, all of these decisions really mattered only if you were a person to begin with. Only persons could hold property. Or take the informed decision not to. In theory, both the Code Civil and the 1844 patent law acknowledged the universal right of holding property. In practice, it applied only to half the population. We have already seen how the Code assigned all rights to the husband and none to the wife. While Marie Curie was not made into her husband's property, her legal status as a married woman made her more like property rather than like an autonomous, rational holder of it. She could work and labor with radium all she liked, but because the law did not recognize her as a person, the universal principles of the Code Civil were a bit less universal when it came to her.

When Le Senne turned to his female readership, trying to explain what they could expect from the law whether they were unwed, married, or widowed, he took extra pains to present his ideas as clearly as possible while pointing out the complexity of the legal language. His was an extremely detailed book, containing brief chapters on women in the army, women as judges of artistic competitions, even, "what a dream, *mesdames*," women ambassadors.[41] As the author of a previous treatise on intellectual property, Le Senne raised the issue also in this volume. However, when it came to women who had secured control of their property via a marriage contract, "one cannot," he wrote, refuse her "the property of her intellectual *œuvres*,"

> works of art, literature or science. But can she freely deal with a publisher for the publication or reproduction of her works? Yes, because this dealing falls under the category of an official deed. For the same reason, she does not need assistance or authorization to take out a patent in her name, register a design or pattern, or concede the right to its exploitation. In one word, she has all the capacity to do all of these things, as long as they do not have a commercial character; if they do, the authorization of her husband is indispensable.[42]

Even if you did not control your own property through a contract, Eugène Pouillet noted that the label of *incapacité* did not apply when it came to the text of the 1844 *loi*. Because there was no pre-examination involved in securing a patent, the law had no business inquiring into the quality or the capacity of the person applying.[43] But if the prejudices of the system could be ignored during the infancy of a patent, they became a reality the day you had to defend yourself against infringement. As Le Senne went on to stipulate, the Code Civil made it impossible for a woman to speak for herself in court. And this was because of the one general, absolute rule, the one without exception; women could not "plead a case in court relative to their person or their property, not as plaintiffs, nor as defendants, without being authorized by their husbands or the court."[44]

Now, if enforcement was the crucial element in the intellectual property system, then it did not really matter if women could take out a patent, because they were powerless in a patent dispute case. By comparison, the U.S. Patent law explicitly included a "she" among the persons who could petition for a patent at the Patent Office. However, while marriage laws had begun to change in the wake of the Civil War, similar exclusion from the courts remained also in the United States. Nonetheless, it has been argued that the cheaper costs involved in applying for a patent as well as the broader base of innovator recruitment made the United States more friendly to innovation than contemporaneous European nation-states. While the number of female patentees was modest, women contributed to patenting in the United States in greater numbers than in France.

And while intellectual property law would remain tied to the nation-state, new international multilateral treaties such as the Paris Convention for the Protection of Industrial Property of 1883 and the Berne Convention for the Protection of Literary and Artistic Works of 1886 initiated a streamlining of intellectual property law across jurisdictions and borders that continues to this day. A revision of the Code Civil, however, was nowhere in sight. Modest changes had been suggested already in 1890, for example a mechanism shielding revenues of employed women from their husbands in case of any wrongdoing on the part of the latter. Jeanne Chauvin—famously refused access to the Cour d'appel in Paris in 1897 despite having all her credentials in place—had to wait three years for the *loi Viviani* to allow women access to the bench. Chauvin was a strong proponent for reform and argued that the contract was the only means by which women could be accorded freedom under the Code Civil.[45] Nothing forbade women to use the means of the contract in order to secure control over their property, but this was a costly affair and mostly limited to the upper classes.

Marie and Pierre Curie never signed a marriage contract, but Irène Curie and Frédéric Joliot did. Dated October 7, 1926, the document goes into great detail about the separation of property between the two spouses. Each conserved their right over their own tangible and intangible property, and the contract gave Irène a wide berth and autonomy to enter into agreements and sign contracts without the consent of her husband. Groomed to succeed her mother, Irène had defended her thesis in 1925, and one can only surmise that the contract was meant to secure her autonomy in decisions related to the running of the Radium Institute.[46]

It is worth lingering briefly on the discourse of family life, partly because of the kinship between the words patent and paternity and partly because the nuclear family is a rhetorical trope in Curie's life and work that remained a constant throughout her career. Recall Paul Lucier's earlier description, framed in all-male terms, of the dilemma of the nineteenth-century scientist, forced to choose between "his professional integrity" and "his rights." Priority claims and patents affirmed science as a cult of masculinity, and the relation between patent and paternity could not have been more clear. This cult of masculinity—which saw the invention or discovery as well as the inventor or discoverer in gendered terms—was underwritten by the law's insistence on reserving the category of citizen and person only to men. The

"sexing" of science took place in print—the various combinations
of authorship, the Madame Skłodowska Curie, the abbreviated
S. Curie, and the final Madame P. Curie versus the stable Pierre
Curie—and in being assigned or stripped of the right to defend
the knowledge produced under these names. It seems perhaps a
far-fetched comparison, but if proving paternity of a patent or
priority in a scientific *note* related to a formal acknowledgment
of parenthood, what about paternity suits in real life? That was
quite a different matter and forbidden by article 340 in the Code
Civil until 1912. As we shall see in the fourth chapter, when the
controversial idea of scientific property came on the international
stage in the 1920s, the paternity search was one of the meta-
phors used to illustrate the problem of ascertaining exactly which
scientist would benefit from protection. That story will have to
wait. For now, the fact that one legal avenue was closed opened
up other doors. Considering that women could not go to court
to defend their intellectual property, it is quite interesting that
in order to establish paternity, or "authorship of the pregnancy,"
women turned to article 1382, which dealt with damages and rep-
arations, shifting, as Rachel Fuchs writes, their "legal narratives
from paternity to property damage and broken contracts."[47]

Marie Curie contrasted patents and publications as the two
poles of a gift/market economy, and the distinctions she made
between them in *Pierre Curie* seem rhetorically all-important.
Yet the strategies deployed from the 1888 *brevet* to the *note* and
from the EPCI laboratory to the industry alliances that began
to form around industrialist Armet de Lisle, who opened a factory for the manufacture of radium salts at Nogent-sur-Marne
in 1904 and that same year helped launch the journal *Le Radium*,
were all part of the Curies' portfolio. The science community and
the industry community were both invested in promoting the
commercial and therapeutic potential anticipated from radium,
but Curie did not always appreciate the business practices of the
radium companies. She had judged the radium ores of the South
Terras Mines in Cornwall weak and difficult to treat. Nonethe-

less, the company had made unjustifiable use of her name and scientific authority in their publications, "a veritable abuse" that enraged her.[48]

I want to suggest that her careful calibration of what "I" and "we" did also reflects a division of labor. The person of the "I" was allowed only some types of ownership, could make only some claims. That she would have patented had she been able to defend the patent in court is improbable. Ultimately, it might even be uninteresting. What is more important is to recognize that the authorship/ownership strategies that were available to Marie Curie in order to be identified as the "author" rather than "owner" of radium developed both inside and outside the intellectual property regime proper. The ownership conferred by "I" and "we" goes directly to the question of the consequences of the Curies' collaborative work in the laboratory and what sort of author and authority it produced. Because the law excluded her from the status of person upon which these intellectual property rights depend, the "property" road was closed to Marie Curie. The persona road was not. And while the law did not allow her to be a person, she was becoming very good at cultivating her persona.

When Marie Curie formulated the ideological statement of disinterestedness in *Pierre Curie* that opened this chapter, she did so from the privileged vantage point of being alone with her story. Discounting her longstanding collaboration with her daughter Irène, she had lived and worked without her original research partner for almost twenty years. In 1923, her fame eclipsed her husband's many times over. She could hardly erase herself completely when accounting for the agency behind their scientific ethos. Instead, she chose words that almost made it look as if her husband had asked her for permission to enforce the principle of nonpatenting in practice. "With my agreement" (*d'accord avec moi*) suggests that the final disinterested word came from Marie and not Pierre Curie. But the rest is even more intriguing; "*he* refused to draw any material profit from *our* discovery." The oscillation between who was actually in control of the intellectual work produced by the couple reveals that "he," not "we," refused to make profit from what was "our" discovery, their common work. Freudian slip or not, she was right in saying that only one person in this collaboration could profit or choose not to profit from their discovery, and that person was her husband. The rue Lhomond hangar gave advantage to claims made in the name of the father, and not the mother, of radium.

IV

On April 20, 1906, *Le Figaro* reported that Pierre Curie, the "inventor of radium," had died the previous day, "victim of a horrific and absurd accident."[49] Eulogizing him as the inventor of a naturally occurring phenomenon was a bit of a stretch, but the finer points of the discovery/invention distinction were not foremost in the mind of the press that day. Crossing rue Dauphine, he had slipped on the wet ground and was run over by a horse-drawn carriage. Newspapers spared no detail of how the scientist's head was crushed under the wheel and held nothing back when recapitulating the dramatic moment when the identity of the victim was revealed.

The accident could not have happened at a more inopportune moment. A mere six months following Marie Curie's June 25, 1903, defense of her thesis, *Recherches sur les substances radioactives*, the Curies received two major accolades; the Davy Medal from the Royal Society and the Nobel Prize in Physics, which they shared with Henri Becquerel. Theirs had been an uphill battle, but suddenly the tide had turned. On October 1, 1904, Pierre Curie finally got his coveted chair at the Sorbonne, and exactly one month later, Marie Curie was named *chef de travaux de physique* in his laboratory, earning an annual salary of 2,400 francs.[50] A year later, Pierre Curie became a member of the Académie des sciences, and the outsider turned into the penultimate insider. In the international public eye the couple were never more of a "we" than in the famous 1904 *Vanity Fair* caricature showing Pierre Curie holding a fluorescent test tube in his hand and Marie Curie standing behind him, her hand on his shoulder.

Ten days after Pierre Curie's death, on April 30, Marie Curie began what is now known as her "mourning journal." It consists of altogether eight entries, five of which are written as letters to her deceased spouse and three of which have a less epistolary tone. In the letter to her husband on May 1, 1906, less than a month following his death, Marie Curie wrote that she had now been nominated to succeed him at the Sorbonne, to teach his course and lead the direction of his laboratory. "I have accepted," she conceded, but then doubt sets in: Was she really qualified? Could she pull it off? "I put all my hope of scientific work in you, and now here I am daring to undertake it without you." Anticipating his reply, she knew that he would have wanted her to continue their work. Turning to him as if he were sitting across from her, she wrote, "how many times have you not told me yourself," and then she articulated what she had heard him say to her repeatedly: "'if you weren't there anymore I might still be working, but I would be no more than a body without soul.'" She ended with a question. "Where shall I find a soul when mine has departed with you?"[51]

This body without soul reappears in the *Dictionary of Scientific Biography*, a first port of call for anyone needing brief biographical data on a scientist. That the entry on Marie Curie relies substantially on the *Pierre Curie* biography is perhaps to be expected, but it is more troubling that *DSB* would take the quotation above and turn it into "whatever happens, even if one were to be like a body without a soul, one must work all the same."[52] Not quite what Curie wrote in the "mourning journal," where there is no neutral and neutered "one," as the *DSB* version would have it. Only *you*. Although this one-stop shop for scientific biography does not list Eve Curie's biography of her mother as one of its sources, what we are reading is in fact a verbatim quote from *Madame Curie*, where the exchange appears in a love scene between her parents at the time they were newlyweds. Marie Curie turns to her husband and asks, "We can't exist without each other, can we?" He answers her, saying: "You are wrong. Whatever happens, even if one has to go on like a body without a soul, one must work just the same."[53]

The simple answer to why the *DSB* did not use the original text but Eve Curie's version is that according to Curie biographer Susan Quinn, the "mourning journal" did not became generally accessible until 1990, many years after the *DSB* was published.[54] But the fact remains; this revisionist scene is an isolated, but telling, example of the extent to which subsequent sources on Marie Curie, such as encyclopedias or dictionaries, were influenced by and completely relied on *Madame Curie* as providing an accurate and truthful account of Marie Curie's life.

A slick pavement, a heavy *camion*; and age thirty-nine, Marie Curie was a widow and single parent to her two young daughters, aged two and nine. Strangely, life went on as before. But it was not only the children who needed her attention. On May 8, a date when there is no entry in the journal that helped her cope with the emotional tsunami of sudden and unexpected grief, one of Pierre Curie's closest friends, the chemist Georges Gouy, wrote to Marie Curie and assured her of his continued friendship and support. He was also quite frank about another, more delicate subject. What was to become of the radium, given the circumstances? The very first thing, he urged her, was to figure out the role of the state. An official inventory would have to be made, signed by the *doyen*, specifying how much the university was in possession of. On such an inventory, he underlined, you *must not mention the radium that belongs to you*, in order to avoid paying inheritance tax. He knew that the Curies considered radium not as ordinary property but as a work tool from which one should not try to make a profit. After a few philosophical ruminations on the current and future value of radium, he continued: "I understand perfectly that you do not want the future husbands of your daughters to require an account of something that has no market value for you." He could hardly have made his point more clearly. Marriage and a sexed legal property regime made radium instable, threatened as a scientific discovery. The risk of losing control over radium was very real for the next generation. Legally speaking, Gouy thought the radium belonged, at least in part, to Pierre. He urged her to seek legal advice on the matter as quickly as possible.[55]

Throughout the coming years, questions pertaining to heritage, succession, and control of the radium the Curies had renounced reappear at frequent intervals. Maurice Crosland once wrote: "any body that controls scientists controls science."[56] He was referring to the work of the Académie, but he could have been referring to the Code Civil. If the death of Pierre meant the loss of Marie's soul, it also gave his widow a new body. And this body, the widowed body, *was* entitled to hold property.

2

Scandal, Slander, and Science: Surviving 1911

Before that fatal accident in the spring of 1906 ended both their marriage and their scientific partnership, the Curies lived a few frenzied years together as international celebrities. And it was the Nobel Prize that turned local renown into global fame. Today, when laureates bring entourages of friends and family to spend the best part of a week lecturing and sightseeing in Stockholm, choosing to abstain from a trip of such magnitude because of general fatigue and teaching commitments sounds like an odd thing to do. And yet, in 1903, the Curies did just that. They politely but firmly declined the invitation. True, the Nobel Prize was only in its third year and the award dinner far from the hyperanticipated event it later became, but it was still a bit of a brushoff. They were too exhausted and too busy teaching to undertake such a long trip up north, Pierre Curie wrote to Christopher Aurivillius, chairman of the Royal Swedish Academy of Sciences.[1] Pierre Curie was never an avid traveler, and his reluctance to leave Paris was perhaps compounded by the miscarriage Marie Curie had suffered in August, five months into her second pregnancy. His wife was still frail. Sweden was a cold place in December. They would try to make the trip in the spring.

On the day the Curies were supposed to accept their Nobel Prize from the hands of the Swedish king, *Le Temps* published a long account of a visit to the home of the laureate duo. Gaston Rouvier braced himself for his trip to the remote house on 106 Boulevard Kellermann almost as if embarking on an expedition to Siberia. When he finally arrived in the outskirts of the thirteenth *arrondissement*, the journalist had to knock on the door three times before the maid opened and told him that Pierre Curie was in the laboratory and Marie Curie in Sèvres, teaching. The prospect of returning empty-handed from such a taxing excursion was enough to convince Rouvier to stay where he was until Pierre Curie returned home. In the meantime, he made do with six-year-old Irène, finding it difficult to take his eyes off the child when she ate "all alone, so that her mother could win the Nobel Prize."[2]

It was worth the wait, of course. Belle époque readers had a voracious appetite for intimate details on stage actors, politicians, authors, and artists, who felt safe enough in their own home to share their feelings and aspirations with the right kind of reporter. Radium had seen to it that now scientists, too, counted among the dignitaries of the day.

I

"A celebrity crystallizes," legal scholar Sheryl Hamilton writes, "when we both desire and recognize a relationship between a person and a persona."[3] Catering to such collective longing was something publishers had been busy doing for some time. In 1788, the *Avant-propos* of *Les illustres modernes* suggested that its unique compilation of one hundred famous Europeans was nothing short of "public education in action."[4] The private life of celebrities was a didactic compass; their accomplishments could be imitated and their mistakes avoided. Of course, such an instruction manual pretty much forfeited its purpose if the public for which it was intended could not afford the cost of learning. Books were just too expensive for the audience of *Les illustres modernes*, which is why it appeared instead as a modestly priced booklet in ten Monday installments, each featuring a lifelike portrait and a two-page biography of ten famous personalities. The bulk of the prerevolutionary roll call was made up of the royals, clergymen, aristocrats, and officers one would expect, but a handful of accomplished *savants* rounded off the representation. Regardless of privilege, those who counted as illustrious moderns lived a life far removed from their readers', but not too far away to lose their function of role model completely. *Les illustres modernes* also made good on its promise to include both men and women among the hundred, even though the final ratio came out a whopping ninety-six to four.

At the time the Curies found themselves involuntary participants in celebrity culture, concepts like status and fame were increasingly associated with real achievement, not just inheritance. Thus, there was a world of difference between being counted as a celebrity by *Les illustres modernes* in 1788 and by *Le Temps* in 1903, and it had everything to do with a seismic shift in the importance accorded to social pedigree. As a couple, the Curies were tailor-made for a republican meritocracy defended and hated in equal measure during the belle époque. Pierre Curie was from a family of freethinkers and famously turned his back on the French establishment when during the Nobel year he rejected the Légion d'honneur with: "I do not in the least feel the need of a decoration, but . . . I do feel the greatest need for a laboratory."[5] On that ill-fated April afternoon in 1906, he had just come from a meeting with the recently established Association des professeurs des facultés des sciences, a new organization formed outside of and in opposition to the traditional old boy's network of the *grandes écoles*. She was Polish, mother of a young daughter, and had recently defended her thesis—the first ever by a woman—at the Sorbonne. They were certainly illustrious, but more than anything, they were *modern*.

However much *Les illustres modernes* in 1788 consciously chose a publication model ensuring the largest possible audience, used portraits as visual proof of genius, and penned the explanatory text to each medallion in the kind of hyperbole that predates today's tabloids, the arrival of the mass press was a definite game changer in the evolutionary timeline of celebrity culture. *Les illustres modernes* had selected and arranged and disseminated, but the mass press did much more than that. Not only did it invent personas better and quicker than its precursors, but it also had the means to posit itself as the gatekeeper of the inner person everybody knew—largely because the same source had told them so—existed behind the persona. The mass press was an information *perpetuum mobile.*

Perhaps Gaston Rouvier would have preferred it if his at-home reportage had revealed the Curies as a bit more human and likable and a little less quirky and stiff, but finding words to describe their unconventional life was something of a challenge. Traditional scientific marriages of the period saw women and men organizing their respective roles in a complementary fashion, perpetuating a gendered lifestyle erected on a distinct division of labor. To some extent, this held true also for the Curies. After all, Pierre Curie was not only his wife's *chef de laboratoire* at the EPCI, but also *chef de famille.* But theirs was a marriage where the boundaries between the *foyer* and the laboratory were less rigid than was the case for many of their friends. Pierre Curie was no more traditionally patriarchal than Marie Curie was the kind of wife whose role was limited to household chores, playing hostess at elaborate dinner parties, or corresponding with wives of other famous scientists. They were a pair of nonconformists conforming.

Eve Curie described with panache in *Madame Curie* how her parents were approached for samples and information by the international scientific community. But it was not all work and no play for the Curies in belle époque Paris. Folies-Bergère sensation Loïe Fuller wanted them to make her a radium dress, and they struck up an unlikely friendship with the American performer. Partly owing to Pierre Curie's avowed interest in spiritualism, husband and wife attended several sessions with the medium Eusapia Palladino, a celebrity in her own right. In what was to become his last letter to Georges Gouy, sent just a few days before he died, Pierre Curie wrote that it was now impossible for him to deny that the phenomena he had witnessed during her séances were genuine.[6] Such faith in Palladino might strike us as sweet if deluded, but he was not the only one who believed her capable of levitating and communicating with the dead. Many of Pierre Curie's scientific contemporaries were every bit as intrigued by the unknown as he was. Perhaps Eve Curie felt that her father's interest in the supernatural would end up an embarrassing stain on his scientific legacy, and in *Madame Curie* she carefully pointed out that her parents were observers, not followers, of the Italian spiritualist.[7]

On the same day as Rouvier's article appeared in *Le Temps*, *Le Petit Parisien* competed for readers' attention with a slightly more spectacular piece, subheaded "United in Science" and "The Husband's Wound."[8] In an uncharacteristically ebullient mood, Pierre Curie showed the reporter an irritated sore on his arm, caused by holding a small vial of radium close to the skin. *Le Petit Parisien* hailed the lesion like a badge of honor, proudly worn as evidence of a personal commitment to science that defied the apparent dangers exuding from the amazing new element. It was as if Pierre Curie had returned from the front, the scientific front, that is, wounded but victorious. Such display of self-sacrifice, duty, and discipline were hallmarks of the French belle époque cult of the hero, a cult by and large reserved for men.

Thus, the Davy Medal and the Nobel Prize were decorations awarded for bravery in war, resulting in the kind of celebrity status that Eve Curie claimed forced her parents to use assumed names when checking into small inns and hotels during vacation.[9] Referring to the two prizes as a watershed moment in their careers, "from then on," Pierre Curie's longtime assistant Albert Laborde remembered many years later, "it was necessary to pay the price of fame."[10] Marie Curie chose "burden" for the heading of chapter 6 in *Pierre Curie*, where she described their strategy vis-à-vis fame as one of reluctant acceptance. But they were no fools when it came to understanding the power of the press, and as much as they abhorred journalistic intrusiveness, the Curies also needed and depended on print media. Perhaps they were even better at using it to their advantage than we might have been led to believe.

In a letter to Georges Gouy in the winter of 1904, Pierre Curie complained about "the stupid life" he was living at that moment. Possibly referring to Gaston Rouvier's visit, he noted the absurdity of a "conversation between my daughter and her nurse" being reported in the news. He ridiculed the autograph collectors, snobs, and even the occasional scientists who came to visit their "magnificent" rue Lhomond laboratory. He was being ironic on that last score, of course, but also candid when he confessed that he could suffer all the noise and then some if it would only give him a chair and a decent laboratory.[11]

Media coverage was precisely the double-edged sword Pierre Curie made it out to be. While the press represented an irritating disturbance of the Curies' work, work they would have preferred to continue in anonymity, it was also the foremost vehicle by which their priority and importance could be affirmed and upheld. And by the beginning of 1904, the Parisian press had made radium and radioactivity *salon* buzzwords and the Curies the talk of the town. Two years later, Pierre Curie's tragic but spectacularly newsworthy death changed everything. Not only did it mark the end of the "we" of an internationally recognized scientific partnership that had become the pride of the French nation, but even more important, gone was the celebrity couple that since 1903 had become sought-after prey for the press. As long as Pierre Curie was alive, Marie Curie was the irreplaceable sidekick, a collaboration curiosity where there was never any real doubt about who led and who followed. Rightly or wrongly, he was the trailblazing innovator of radium: she the enabler, the shoulder of support, the capable assistant. Widowhood meant that new lines had to be drawn between Marie Curie and the press, borders that demarcated an unknown territory. There were familiar enough images and words to describe *him*, but someone like her had not really existed before. Marie Curie *sans* Pierre was a clean slate.

In the previous chapter, I briefly noted the intimate rapport between the press and scientific institutions and scientists at the time of the discovery and isolation of radium, one that included information as well as people. François Roulin, the working editor of *Comptes rendus*, had previously been employed by *Le Temps*. Gauthier-Villars, the publisher of *Comptes rendus*, also published Jules Verne. Hardwired into the system, the Curies relied on a bundle of print options in order to claim priority and control of their expertise as well as of their radium. In this chapter, I want to zero in on the year 1911, Curie's personal *annus horribilis*, and consider how her relationship with the ubiquitous creator of celebrity—the mass press—at this time took a new and dramatic turn.

The events I consider next confine themselves to a few weeks in the beginning and end of the year, but they are neither isolated from the prehistory covered so far, nor detached from the legacy about to form. Bookended by the failed candidacy at the Académie des sciences in January and the drama of the Langevin *affaire* in November–December, 1911 really is a landmark year in understanding how Marie Curie's public persona consolidated into that of a famous celebrity outside the marriage that had defined her both scientifically and legally. Eight years after she was catapulted into fame by the first Nobel, the composite weight of the 1911 events, the burden of celebrity, resulted in five duels— four with *epée* and one with pistols—fought on November 23, 24, 26, 28, and December 19, 1911. All five involved journalists, and all five were over Marie Curie.

II

Everything began almost exactly a year earlier. In a letter to *Le Temps* dated November 18, 1910, Marie Curie confirmed that the speculations circulating around the capital were correct. She was indeed a candidate for the Académie seat left vacant following the recent death of chemist-physicist Désiré Gernez.[12] The institution in question consisted of eleven sections divided into two main groups—*sciences mathématiques* (5) and *sciences physiques* (6)—each with six resident members (elected for life) and presided over by two perpetual secretaries.

As we saw previously, Curie's relationship with the Académie was a longstanding one. The discovery of radium and polonium had first been announced in the *Comptes rendus*, and she was a multiple recipient of its Prix Gegner, in 1898, 1900, and 1902. But becoming a member of the venerable institution was something quite different from publishing in its journal or receiving its awards, and why she chose this particular time to put her name up for consideration remains unclear. More certain is that Pierre Curie's love-hate relationship with the Académie—he was ambitious for a seat but cringed at the work it took to land one—would have made her a highly reluctant candidate unless she felt there was a real chance of success. As 1910 drew to a close, several developments may have convinced her that despite everything, the impossible was perhaps now possible after all.

As we saw in the previous chapter, only a month following her husband's death, Marie Curie succeeded Pierre Curie as professor of physics at the Sorbonne. It was not unheard of for widows to continue the work of their husbands in earlier guild structures, but the strict division of labor in the new science era abided by a different set of principles. Perhaps it was the collective shock of having one of France's most promising young scientists killed in his prime, and the daunting prospect of losing the French radium advantage because of it, but the academic establishment concluded that the only one who could pick up the pieces without loss of momentum was his wife.

Four years later, the Radium Institute had been established and was slowly becoming the kind of laboratory her husband desired but never got. Step by step, Curie was positioning herself as the central node in a radium network that extended from Paris around the world. The International Congress of Radiology and

Electricity in Brussels in September 1910 was a critical milestone in that project. She had been uncharacteristically aggressive at the conference, campaigning to have the newly established radium standard deposited at her laboratory rather than at the customary Bureau international des poids et mesures at Sèvres. Given the reluctance to patent, hers was a somewhat unexpected proprietary stance that would have allowed her to exercise control over the uses of the standard, for instance through licenses. She lost that particular battle, but won the war. The illustrious Commission internationale des étalons de radium assigned the task of developing the standard to her. More important still, it was decided that it would be known as the curie.

The Curies may have disavowed patenting and spurned commercialization of basic research, but the value of the name Curie was incalculable. Officially the choice honored her dead husband, but she was the custodian of the family name. Years before the events of chapter 4, including Curie's leadership of the bibliographic initiatives of the CICI, the standard was another piece in what was to become the international organization of information and knowledge after World War I. The standard was in line with a two-pronged approach that disinterestedly left radium alone but protected the Curie name, very much in the same way that a brand would be protected and cared for. There was no way the name could be mentioned without reminding people of the husband-and-wife team, the flesh-and-blood persons behind the discoveries. Not the eponym of an era, but close enough. As a standard, the curie also belonged to the same depersonalized and universalized category as the meter, providing the ultimate proof of French rationalism and internationalism. It really was a win-win situation.

Marie Curie had claimed many firsts. Now she aimed for the ultimate one, as inclusion among the *immortels* of the Académie would make her the first *immortelle*. At first blush, her radium track record should have made her a shoo-in for Académie membership. And yet hers was a monumental aspiration, the consequences of which needed to be downplayed. In her brief letter to *Le Temps* Marie Curie adamantly stressed that election particulars always were kept out of the public light and that she would find it embarrassing if this custom would in any way be modified because of her. In light of what she knew of the power and size of the French press, her wish sounds oddly naïve, if not slightly coquettish.

The Parisian press circulated information, created and destroyed celebrities, and acted as political platform across the left-right spectrum, all on a massive scale. In November 1910 and 1912 the total print runs for the daily papers were 4,950,000 and 5,270,000. Paris had around seventy daily papers, but the market was dominated by the four great ones whose 1912 print-runs put them in a league of their own; *Le Petit Journal* (835,000), *Le Petit Parisien* (1,400,000), *Le Matin* (670,000), and *Le Journal* (810,000). In 1914 their combined circulation constituted almost three-quarters of that of all Paris publications and 40 percent of the French total.[13] Regional newspapers had experienced an impressive growth, with print runs multiplying by six between 1880 and 1914, and there were seemingly endless additions to the specialist press.[14] Sold as a daily for one sou, *Le Petit Journal* (1863) had led the way for the mass press in France, and innovations in technology (*Le Petit Journal* adopted the rotary press in 1867), education (compulsory schooling arrived in 1882), and legislation (notably the press law of 1881) created an infrastructure for the phenomenal impact of the mass press.

In early December, *Le Figaro* added to the melee with Abel Faivre's caricature of a woman wearing the famous Institut building on her head and with a caption that read: "what a beautiful hat the Cupola will make!" The text featured below the sketch was less flippant in tone; it was a report on a meeting held by the administrative committee of the five autonomous academies that together constituted the Institut de France.[15] Now that Marie Curie had officially come knocking on the door, the committee had convened to discuss how the Institut as a single body should handle the conundrum of female candidates. Present were twelve members, who "despite their customary urbanity" exited from a session that lasted several hours "as friendly as prison gates!"[16]

From now on, everything focused on one thing: preparing for the upcoming trimestral meeting of the Institut, where the question of female membership would be put to the vote.

On New Year's Eve 1910, *Le Temps* published a long letter from Gaston Darboux, one of the permanent secretaries of the Académie, where he listed the arguments in favor of Curie's candidacy.[17] Interestingly enough, Darboux had been one of twenty Académie members who had nominated only Pierre Curie for the 1903 Nobel Prize, omitting completely any mention of Marie Curie's contributions. It was only through the intervention of Swedish mathematician and member of the Royal Swedish Academy of Sciences Gösta Mittag-Leffler—who took it upon himself to alert Pierre Curie of the embarrassing oversight—and a bit of juggling with the nomination rules, that Marie Curie's name was added to that of her husband.[18] Perhaps Darboux felt that standing firmly in her corner now atoned for his faux pas then. In his attempt to sway readers of *Le Temps* that Marie Curie had never been only "the auxiliary of her husband," Darboux relied on the highest possible authority, Pierre Curie's own words from his 1903 Nobel speech. Having dispensed that crucial ex post facto reference, Darboux was free to expand on Marie Curie's *independent* achievements (isolation of pure radium, the publication of the two volumes on *radioactivité* in 1910, and countless memberships in learned societies worldwide), adding finally that because she was the undisputed leader in the booming field of radioactivity, the Académie really needed her knowledge and expertise to judge the merits of others. Only thus could continued French leadership in the field be secured. As champagne corks popped on the stroke of midnight and 1910 turned into 1911, the pros and cons of admitting a woman to the famed Cupola for the very first time must have been the fodder of many impassioned dinner conversations.

>>><<<

On January 4, 1911, 153 *immortels* showed up to vote on the question of the eligibility of women into their elite circle, a record number and more than twice as many as were usually present at the trimester meetings of the Institut.[19] The day before, *Le Matin* had asked a few members for their personal views on the matter before them. Émile Roux, director of the Institut Pasteur and one of Curie's supporters, simply stated: "I don't believe that discoveries have gender. I will always support the person doing the best work." Émile Amagat, the physicist who beat Pierre Curie to an Académie seat in 1902, was in a less generous mood. At the last meeting of the physics section, Amagat explained to the newspaper, there had been a vote on whom to suggest as Gernez's successor. Then he revealed that while three of the altogether five members placed Marie Curie first, one voted in favor of Eduard Branly, professor of physics at the Institut Catholique in Paris, and the fifth and final vote, his own, was an abstention. As women could not be members of the Institut, Amagat considered the whole process illegal from start to finish, and he told *Le Matin* that his only option had been to refrain from voting altogether.[20] On the eve of the vote, the general feeling was that the majority would uphold the principle of the ineligibility of women, defending the status quo.[21]

And so they did. On paper, at least, the result of the two questions set before Institut members seemed to go against Curie. The first vote went in favor of letting each Académie decide on the eligibility of women for themselves. The second and more ambiguous outcome was that with 88 votes against 52, the Institut considered the ineligibility of women a tradition it would be wise to follow. There were varying interpretations of just how closely this advice ought to be heeded, though. Gaston Bonnier of the Académie referred to the recommendation, tongue-in-cheek, as the kind of advice friends sometimes make, "knowing it would not be taken."[22]

Things might still have gone Curie's way. *Le Matin* reported that a recent meeting of the nominating committee ended with Amagat, that "irreconcilable enemy of female candidates," storming out of the room shouting, "I'm completely demolished!"[23] When the physics section a week later presented their official list of candidates in the *Comptes rendus*, Marie Curie once again came in first. With the final vote scheduled for January 23, the announcement revealed that she was running against six male competitors. As expected, Eduard Branly proved the most qualified in the group, and Eve Curie described how the press did their outmost to exploit the election as a battle between the "Curistes" and the "Branlystes."[24]

A famous *Excelsior* cover from the period gives some indication of the way in which illustrations and images increasingly figured in the daily press. The double visual depiction of Curie—a somber full-face portrait and a smaller profile—was accompanied by another illustration, a reproduction of one of her letters. It was all a tantalizing lead-in to the expert opinions provided on the "physiognomy and handwriting of Mme Curie." Madame Fraya could attest that the handwriting revealed a refined mind, a loyal and reliable character. Madame Lioubow felt that the nose might be a bit too short in respect to the face as a whole, but noted a "bold attitude matched by a kind of careful deliberation." Perhaps some readers felt reassured by Mesdames Fraya and Lioubow's mutual graphology/phrenology conclusion that Curie's love of science went hand in hand with a discreet feminine sensibility.[25]

Not everybody was convinced. Indeed, many felt that what was at stake in the choice between Branly and Curie was much more than the question of whether a woman should be accepted into the Académie. The choice was between two worlds, not two sexes. *Le Petit Parisien* wrote that Curie had the backing of her colleagues at the Sorbonne and "academics taken with modernism." Branly's sponsors, however, emphasized the "high scientific value of their candidate, who is really a scientist appreciated throughout the world." There could be little doubt that a vote for Curie was a vote for emotional turmoil and coat turning, whereas a vote for Branly meant recognizing universal values and tradition.[26]

As the election date drew nearer, *L'Excelsior* relied once more on its favored photocollage technique and placed both scientists in a scale, weighing equal, image-wise, in the race for the Académie. In the accompanying text, however, Curie again found her accomplishements in a secondary rank, having only "collaborated with her husband in the discovery of radium," whereas Branly's contributions, as "inventor of wireless telegraphy," indicated the kind of autonomy and agency associated with independent scientific work.[27]

An obvious favorite with Catholic newspapers such as *La Croix*, Branly was compared with Curie in terms of seniority as well as academic independence. *La Croix* referred to his abysmal working conditions, in stark contrast to the marvelous resources "Mme Curie benefits from," leaving out her earlier history of insufficient funds and laboratory space. She was a neophyte, but Branly a repeat candidate, and *La Croix* felt strongly that tenacity and age were overdue for their just reward. And what was, in effect, they asked their readers insidiously, Marie Curie's "*actual* contribution to her husband's discoveries?" devaluating her involvement while simultaneously conferring all authorship status for radium on Pierre Curie. While he was still alive, *La Croix* trumpeted, France could defend a first place in all things radioactive. Now, after his death, "the English, with Rutherford, Ramsay, Sody [*sic*] and several others, have left us behind." In other words, a woman at the helm meant irreparable harm to the preeminence of French science. Branly, on the other hand, had fathered that wondrous invention *télégraphie sans fil*. When the Royal Swedish Academy of Sciences awarded Guglielmo Marconi and Karl Braun the 1909 Nobel Prize in Physics, *La Croix* could not have been more incensed; it was a reward for TSF from which, "incomprehensibly, Branly was systematically excluded!"[28] There was only one possible explanation: Branly was a victim of an ongoing anti-Catholic conspiracy on the part of the academic establishment, one that only Académie membership would redress once and for all.

And in the end, Curie lost to Branly by two votes, 28 against 30. *Le Matin*'s January 24 front page featured a photo from the meeting and a commentary concluding that choosing Branly had been a gesture of justice and patriotism.[29] Even a pro-Curie newspaper like *Le Temps* acknowledged that it would have been unfortunate had Branly been rebuffed a third time.[30] Elated or disappointed, regardless of how the press viewed the outcome, they were only too happy to feature accounts on how a tumultuous crowd fought to gain entry to the meeting room only to endure an excruciatingly long and boring wait until four o'clock, when the regular proceedings were over and the voting finally took place.[31]

Autopiloted for months, the rumor mill began churning again in the days following Curie's defeat, predicting that it was only a matter of time until she would be in a position to launch her next candidacy. The death of mathematician Jules Tannery presented a new opportunity, even if it meant an opening in the wrong group, *sciences mathématiques* and not *sciences physiques*. News from the Palais Mazarin was cautiously optimistic, and *Le Figaro* outlined in detail the steps that were involved should she wish to retry.[32] She never did. Actually, it would take sixty-eight years until Yvonne Choquet-Bruhat in 1979 became the first woman elected to the Académie des sciences. Whether Curie's decision was brought on by hurt pride or a sense of self-preservation is impossible to say. More certain is that she could not ignore this incident in the "autobiographical notes" she had agreed to write for the U.S. edition of *Pierre Curie*. Somehow, she had to give her side of the story.

So, in 1923, when Marie Curie looked back on this period in her life, she wrote that it was her "strong distaste for the personal solicitation required" that made her decide against reapplying. Fair enough, but there was certainly nothing odd in trying a second or even third time before landing that coveted seat. Others before her had. Her husband, for one, swallowed his pride and repeated the visits, the handshakes, and the smiles until he was rewarded with admission on his second try. Paul Langevin tried *four* times, in 1921, 1922, 1923, and 1927, before finally getting elected in 1934. But more than a decade after the fact, Marie Curie was still disappointed, possibly angry; both at herself, for not taking the resistance she was up against seriously enough, and at the Académie, for being such a hopelessly retrograde establishment. As opposed to the way she highlighted her own role in the strategy of disinterestedness, this time others, not "I," made things happen.

> At that time also, several colleagues persuaded me to be a candidate for election to the Academy of Sciences, of Paris, of which my husband was a member during the last months of his life. I hesitated very much, as such a candidacy requires, by custom, a great number of personal visits to Academy members. However, I consented to offer myself a candidate, because of the advantages an election would have for my laboratory. My candidacy provoked a vivid public interest, especially because it involved the question of the admission of women to the Academy. Many of the Academicians were opposed to this in principle, and when the scrutiny was made, I had a few votes less than was necessary. I do not ever wish to renew my candidacy, because of my strong distaste for the personal solicitation required. I believe that all such elections should be based wholly on a spontaneous decision, without any personal efforts involved, as was the case for several Academies and Societies which made me a member without any demand or initiative on my part.[33]

The draft version of the text above shows that she first wrote "I decided," but changed her mind and substituted "I consented," shifting the initiative away from any personal ambition to her colleagues' powers of persuasion. But the most telling editorializing is when she first acknowledges in the draft that she had a few less votes than her "masculine competitor," a fact she deletes altogether in the final version, where she refers to a "few votes less than was necessary." Bitter or not, in 1923 she had enough honors to compensate many times over for that annoying defeat, and she contrasted the distasteful personal solicitation associated with a successful Académie candidacy with her preference for the "spontaneous decision" of the "several Academies and Societies which made me a member without any demand or initiative on my part." Maybe there was a thorn in her side after all, because in the draft version she started out with "considerable number," then modified to "many," and ended up with "several." It was

important not to let interest overshadow disinterestedness *too*
much.[34]

III

The bimonthly women's magazine *Femina* contains the best evidence of just how far Curie's celebrity status extended at the time
she sought membership in the Académie. Launched in 1901, *Femina* catered to a modern, urban readership, and its editorial policy rested on a combination of news, gossip, and human-interest
stories with an emphasis on stereotype-breaking pursuits.

Few women were as stereotype breaking as Curie, and few
periods in her life proved it so conclusively as her 1911 experience with the Académie. The stakes had never been so high, the
investment never so great, and the failure never so humiliatingly
public. And *Femina* covered it all in pictures of Curie hounded
by paparazzi outside her laboratory, trying to shield herself from
the cameras with her bag. The text referred to the recent controversies over her candidacy and noted that she was "involuntarily
and despite herself a person in the news. Journalists and photographers relentlessly assault her home and laboratory." And while
Femina coyly took pains to disassociate itself from this particular
photographer, claiming he was not on its payroll, it was also reluctant to withhold from its readers the fascinating snapshots
taken outside the door of her rue Cuvier laboratory.[35] "Fascinating snapshots" taken by "merciless photographers" revealed much
of the same kind of invasive presence we associate with present-
day celebrity hunting.

Owing to the mediating influence of magazines such as *Femina*,
by early spring 1911 Marie Curie was more of a celebrity than ever
before. This time, however, she was not the co-discoverer, collaborator, or co-anything, but candidate for a powerful academic
institution, *alone*. There were notorious precursors in French
society; George Sand in her time, and, of course, larger-than-
life Sarah Bernhardt, whose celebrity status far surpassed that of

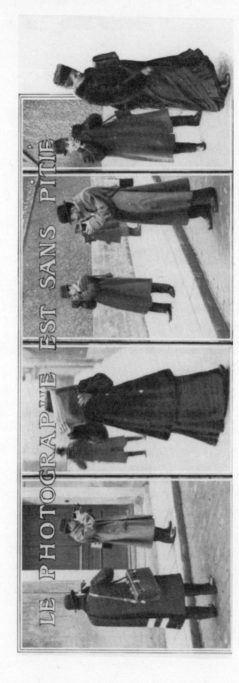

FIGURE 2. "Le photographe est sans pitié" (The photographer is without mercy). Paparazzi taking pictures of Marie Curie outside her laboratory following her unsuccessful candidacy to the Académie des sciences in January 1911. *Femina*, February 1, 1911. With permission from the Musée Curie (Association Curie et Joliot-Curie), Paris.

Curie. Eccentrics like Sarah Bernhardt were accepted, embraced, and even beloved as *eccentrics*, a status that conferred a "hyper-individuality" outside normal rules of accountability. This was not Curie's kind of celebrity, which instead embodied the kind of womanhood sanctioned and admired by *Femina* and its readers. Her black dresses and grave demeanor epitomized the celebrity of "ordinariness," not the radical individualism displayed by Sarah Bernhardt. Marie Curie was different, but as she was about to learn the hard way, she was not different enough to avoid being held accountable for her actions.

>>><<<

Sometime during the summer of 1910, Marie Curie's longstanding friendship with Paul Langevin, a close friend, colleague, and former student of Pierre Curie's at the EPCI, matured into romance. Married with four children but estranged from his wife Jeanne, Langevin met Curie regularly at a rented pied-à-terre close to the Sorbonne. Around Easter 1911, there was a break-in at the apartment, and letters exchanged between the two lovers were stolen from a locked drawer.[36] It was Henri Bourgeois—not only Jeanne Langevin's brother-in-law but also an editor at *Le Petit Journal*—who made Marie Curie aware of the disgrace that would result if the letters ever reached the press. For several months, the threat of disclosure hovered over her, ominously. And then Fernand Hauser of *Le Journal* decided to pay a visit to Paul Langevin's home in Fontenay-aux-Roses and publish his story on November 4, 1911. To a significant degree, the story revolved around Hauser's conversation with Jeanne Langevin's mother, who was more than willing to provide the journalist with an emotional story of her daughter's imperiled marriage. She told Hauser that Langevin and Curie both had left Paris, and that the incriminating letters were in her family's possession, irrevocable proof of the sordid affair.

The next day, *Le Petit Journal* jumped on the bandwagon, claiming to have known of the affair for quite some time. Continued discretion could have been an option, they told readers, had it not been for the fact that both scientists now seemed to have abandoned Paris, one turning his back on home and family, the other renouncing her books, her laboratory, and her *gloire*. As *Le Petit Journal* knew well, Langevin and Curie's whereabouts were hardly a secret. Both were in Brussels, attending the first of the famous Solvay conferences. Jeanne Langevin confirmed that she was about to seek a separation from her husband, not because of any infidelity, for she was convinced he would see the error of his ways and return to her, but because he had taken their two sons with him. He had hit her on two occasions, and his relationship with Marie Curie made her life insufferable. It had all been endured for the sake of her children, and she was prepared to reconcile if only he came home again. *Le Petit Journal* then reprinted a response from Paul Langevin to what he called the "odious and grotesque slander," confirming from Brussels that he had left his "very jealous" wife three months ago, and responding to her charge of domestic violence by recapitulating an incident when she had struck him on the head with a bottle.[37]

For her part, Marie Curie addressed the rumors in her preferred mouthpiece *Le Temps*. She confirmed that she was in Brussels "at a scientific meeting of the greatest importance," and *Le Temps* backed her up by interviewing her laboratory staff, who provided a detailed day-to-day account substantiating everything she had said.[38] The media avalanche not yet moving at full force, there is nothing defensive about Curie's position. Instead, she took the moral high ground and used *Le Temps* to issue a direct threat: "Henceforth, I will rigorously pursue all publication of writings that are attributed to me or that make tendentious allegations about me. As I have the right to, I will ask for considerable damages that will be used in the interest of science."[39] Recognizing a sensation when they saw one, the press willingly reprinted her warning. Then, Fernand Hauser suddenly had an epiphany and tried for damage control, sending Marie Curie a letter in which he apologized profusely. "I cannot understand," he wrote, referring to his article, "how at that moment, the fever of my profession could have made me commit such a detestable act."[40]

Hauser's apology and Curie's threat of legal action allow me to return to the Curie name, one of the more important aspects of this whole incident. One explanation for Hauser's distress could be that he realized (or had been made to realize) that he was the reason why two world-famous and very influential French scientists found themselves embroiled in a potential scandal, the repercussions of which could be enormous. But even more important, there is also a third scientist involved. And for a dead man, Pierre Curie had a very real presence. We saw earlier in this chapter how Curie pursued a very careful strategy with respect to the naming of the curie standard. It would have looked like a failure if she could not have attached the family name to something so disinterested as the radium standard. As we follow her into the next chapters we will see the development of a distinct, certainly not articulated, but unmistakably brand-ish understanding of the importance of the name and the uses to which it should be put. It is not inconceivable that these strategies had something to do with the events that were about to unfold. For what was at stake here was the honor and reputation associated with the name Curie. And the integrity of these categories was everything, both privately and professionally.

Paris was not only awash with newspapers, but also a city where the press was known for being able to say almost anything with impunity. The press law of July 29, 1881, was intended to free the press and promote public discussion, but ended up creating fertile ground for the circulation of slander, defamation, and gossip. Curie's problem was perhaps further complicated by the way she seemed to fall between the two categories of persons that were subject to defamation: the public persons of article 31 and the private persons of article 32. Damages were considerably higher for public persons. Curie was accused of having an affair with a married man, causing the breakup of a (French) marriage. As long as it was kept outside the press and within the bounds of private life, French society had seen worse. Out of sight meant out of mind. But Marie Curie could not hide. She was very much a public figure, and in her frontal attack on her detractors, clearly stated that any possible damages would be used in the "interest of science." She did everything to present herself as a scientist and to classify the rumors and innuendo as defamation of the scientist rather than the woman. However, while article 31 considered defamation against a broad spectrum of public professionals, from lawyers to politicians, if the offense in question concerned private life, it was simply treated as article 32. The question was how to interpret Curie's existence on the border between these two spheres. In any case, pursuing her threat all the way to court would expose her to public scrutiny and scandal.[41] Weak libel laws meant that under the circumstances, Marie Curie's threat of damages was more a figment of the imagination than an actual possibility.

Despite Hauser's contrition and Curie's threat of legal action, by November 6, the *affaire* was not limited to *la presse d'information* but on the brink of being fully exploited by *la presse de doctrine*. In this second category, one newspaper in particular propelled events toward their dueling climax, and that was the right-wing, nationalist, royalist, and highly anti-Semitic *L'Action Française*, official organ of the ideological and cultural movement Action Française. Launched in 1898, Action Française resulted from an event that polarized the nation: the Dreyfus affair. The controversy surrounding the Jewish officer, convicted of treason and sentenced to isolation on Devil's Island, still reverberates through French society. Action Française's allegiance was never in doubt; the organization was anti-Dreyfus to the core. Many in the scientific community sided with Dreyfus, but Jean Perrin was the only one among Curie's closest colleagues who signed the first petition for a revision of the trial.[42] It is not surprising that Marie Curie's name is missing from such an appeal, but Pierre Curie's absence is more bewildering. As his biographer Anna Hurwic also notes, there is no trace of his being actively invested in this major chapter in French history, and once again, we are left with secondary sources telling us that the Curies were Dreyfusards.[43]

In January of that same year, *L'Action Française* had joined forces with *La Croix* and *L'Intransigeant* in backing Branly's candidacy to the Académie. To Léon Daudet, editor-in-chief and one of the central characters in the events that follow, the choice between Curie and Branly was nothing less than a choice between *Dreyfus* and Branly. Those *imbéciles* who believed the Dreyfus affair buried and done with were mistaken, he wrote with his trademark pomposity. And when he spoke of the epic struggle between "the national soul and the foreign demon," more alive than ever and penetrating all walks of life under a thousand different forms, it was clear that Daudet assigned Marie Curie the role of the demon.[44]

The tradition and honor of the monarchy and military were the only bulwark against the contamination of blood and the threat to the French nation from the Jewish cabal. The xenophobic and anti-Semitic extreme fantasies of the right revolved, as historian Ruth Harris so well puts it, around the twin poles of defilement and purity, clad in syrupy sentimentalism and euphoric rage.[45] The feverish tone might have sought to compensate for the fact that the circulation of *L'Action Française* could not begin to touch those of the four major newspapers and that it remained in constant deficiency mode thanks to an unwillingness to accommodate commercials and advertising.

Ten months after the Académie calamity, Léon Daudet had an old enemy to attack but a new score to settle. Science in general and the Sorbonne in particular were targets for his first contribution to the escalating press coverage of the Curie-Langevin *affaire*. Neither Curie nor Langevin is mentioned in his piece from November 6, but there are plenty of indirect hints embedded in a scorching critique of Science "with a capital S." You are mistaken if you think science and virtue go hand in hand, Daudet argued, because "afflictions of the heart are extremely frequent among scientists" (Daudet used both the male and the female form of *savant* just in case the inference would escape readers). The laboratory, Daudet pounded away, is a site for emotion and adultery, and any idea that the sciences confer virtue is just a hypocritical lie passed on from Protestant countries via romanticism.[46]

Mid-November, the press innuendo (and the pressure from the Curie camp to put a lid on things) compelled Jean Dupuy, president of the Paris Press Syndicate, to make a telephone call to the leading newspapers asking them to impose a voluntary censorship on all comments on the Curie-Langevin affair. He did not call *L'Action Française*. This was a shame, Léon Daudet wrote, because he would have liked to tell Dupuy exactly what he thought of such blatant censorship.[47] From now on, *L'Action Française* noticeably upped the ante. In a daily column entitled "Pour une mère," Maurice Pujo, one of the founding members of Action Française and leader of the organization's street gang Camelots de Roi, came to the defense of the wronged French wife and mother Jeanne Langevin, ranted against the Press Syndicate, and detected a Sorbonne conspiracy on Curie's behalf against Madame L.'s French Family Values. Cleverly capitalizing on the feuilleton form to ensure that readers kept reading (and buying) the newspaper, *L'Action Française* used the generic conventions of the mass press to pit two women against each another for the sympathies of a readership increasingly made up of women.

And while women were assigned a crucial role in shaping French civilization and culture, they were also a threat from within, with the mature woman an especially dangerous creature. The Langevin affair spiraled out of control partly because it erupted during a period of heightened gender anxieties; owing to the rise of nationalism in Germany, France faced a crisis of masculinity *L'Action Française* did its best to tap into. The advice Paul Langevin received from *L'Intransigeant*, questioning his manhood and ability to control the situation, was typical.[48]

Chaos of this kind was to be expected, Pujo stressed, considering that all in the Curie camp were "old acquaintances from the Dreyfus affair."[49] Marie Curie was suffering the backlash from being a foreigner, a woman, and a scientist. Whether or not there would be a complete meltdown of her reputation depended on a few letters, the content of which was still unknown.

IV

Not for much longer, however. On November 23, the letters became public, but not through any machinations of *L'Action Fran-çaise*. Instead, Gustave Téry, according to Robert Reid "an unpredictable, aggressive little man,"[50] used his weekly paper *L'Œuvre* to add his own take on things, publish the full text of Jeanne Langevin's official claim against her husband (due in court on December 8), and, crucially, divulge significant (but no doubt carefully selected and edited) sections of the letters between Curie and Langevin. The most damning by far was one written more than a year earlier, where Curie discussed Langevin's marriage in detail and gave him advice on how to disassociate himself from a liaison she felt threatened his future scientific career. Curie demanded a presence in Langevin's private and professional life; she gave advice as a colleague, friend, lover, and leader of a laboratory. But it was her warning to Langevin not to reconcile with his wife and get her pregnant that really sounded like treason in the ears of the nationalist movement.[51] At a time when birthrates were plummeting and Germany was growing more threatening by the day, low birthrate was a particularly sensitive problem in France, where the population stagnated in comparison to both Germany's and Great Britain's.

L'Action Française would never let Curie forget that she was the widow of a man with a "glorious" name, one that unfortunately filtered down indiscriminately to survivors who did not deserve it.[52] It is true that Curie in some sense left her shadow presence vis-à-vis property and the law and became more of a legal person as a widow, but it is also true that the status of widowhood assigned her a continued place in her husband's shadow. As late as 1956, on the fiftieth anniversary of Pierre Curie's death, Marie Curie was still remembered first as "the lonely widow," then as "the mother," and only last as "the physicist."[53]

But in 1911, because she had tried to enter the Académie by running against the far more worthy Branly, and then tried to lead and give advice to a married man, *L'Action Française* felt convinced Marie Curie had defiled the name of her dead husband, a name she carried more or less by accident. As a consequence (deflecting any possible libel suit), the newspaper never mentioned her by name, but only referred to her as *l'étrangère* or Madame X. In the eyes of her detractors, the letters constituted evidence that Marie Curie was about to abandon her established role as devoted widow and mother for that of public figure in her own right. Curie had, Téry claimed, "deliberately, methodically and scientifically applied herself, through the most treacherous advice, by the most vile of suggestions, to detach Langevin from his wife and to separate his wife from her children."[54] *L'Action Française*, now telling their readers they had been in possession of the documents for more than two weeks, noted with some satisfaction (and a whiff of professional envy) that *L'Œuvre* had disappeared in the morning from all the kiosks and that the rumor in the evening was that the issue had been seized by the police.[55]

Private correspondence was perhaps the one thing considered off-limits for the press. Only a few years later, in 1914, when the Langevin *affaire* was water under the bridge and France teetered on the brink of war, the publication in *Le Figaro* of a private letter (part of a slanderous campaign by *Le Figaro* against the left-wing politician Joseph Caillaux that had been going on for weeks) prompted Caillaux's wife Henriette to enter the offices of Gaston Calmette, the editor of *Le Figaro,* and shoot him point blank. The resulting trial was heralded as the trial of the century. The incident proved not only the depths to which the French press had sunk, but also how limited the public role of women remained. Because she was a woman, Madame Caillaux's deed could be explained by emotional chaos, temporary insanity, and the madness brought on by having to defend the family name. Public display of emotion was perfectly *comme il faut* for a woman. Henriette Caillaux's guilt was beyond doubt, and the only way her defense team would secure her freedom was to portray her as a victim of uncontrollable impulses, the irrational being of the Code Civil. They were successful, and she was acquitted both of voluntary homicide and any premeditation.[56]

Marie Curie was guilty of something much worse. Her transgression was that she, as Téry wrote, had applied herself *scientifically* to lure Langevin away from his family, and her reasoning and planning overstepped the very narrow gender role prescribed to her at the time. Even more damning in the eyes of *L'Action Française* was that her conduct was un-French. "What is not French," and "that which first will solicit surprise, and then indignation in people from our country," Maurice Pujo noted, is that "there are no sincere cries, always excusable, but cold reasoning."[57]

Hardly surprisingly, Léon Daudet was the addressee of the first challenge to a duel. Not the first or the last time; when toward the end of his life he took stock of his dueling encounters as a journalist, he ended up with a tally of fourteen.[58] His inflammatory style had made him a ton of enemies, one of whom was Henri Chervet at the newspaper *Gil Blas*. The upshot of the duel was an exchange between the two journalists triggered by a piece Daudet wrote on November 17. In the article, directly addressed to the politicians Alexandre Millerand—whose legal counsel Curie had retained—and Raymond Poincaré, Daudet claimed that while he believed that the separation between private and public life should be upheld at all costs, he was now compelled to come to the rescue of Jeanne Langevin, a woman held hostage by "deceit dressed in science."[59] Chervet responded by criticizing Daudet's journalistic practices as unethical,[60] an angle he later elaborated when he told Daudet that any journalist "who threatens a woman, under any pretext, is a cad."[61] A few days later, Chervet sent his seconds to the seasoned duelist demanding a retraction or reparation through arms. They were to meet at the Great Ferris Wheel, erected for the Exposition universelle in 1900.

But how was it possible, in late November 1911, for two journalists to choose dueling as a means to settle, well, exactly *what* score? According to Bruneau de Laborie's popular *Les lois du duel*

(1906), the object of the duel was to resolve, definitively, disputes where honor was at stake, when all other solutions had proven impossible.[62] From our vantage point a century later, everything about the five duels seems slightly bizarre. Most of us would have expected the duel to be extinct in the beginning of the twentieth century, but it remained both firmly ensconced in public consciousness and a surprisingly popular practice. So popular, in fact, that dueling accelerated rather than abated in the period leading up to the Great War. It was almost as if a proactive resurgence of masculine tradition and honor could somehow act as a barricade against the impending hell. The *point d'honneur* we recognize from Hollywood movies reaffirmed the stability of social hierarchies and the role of the individual within them. And such a ritualized act, the apex of the cult of masculinity, was a lifeline to an honorable past in a time of unforeseen and rapid change.

Remarkably resilient and adaptable throughout its long history, in the seventeenth century the duel had been the domain of the aristocracy, university men, and the military. Three centuries later, it offered a solution also for civil servants and journalists, all of whom were extremely quick to respond to any perceived insult with a challenge to a duel, by either epée or pistols. And duels involving journalists were among the most common. In Italy, over one-third (34 percent) of all duels fought between 1879 and 1889 were over insults published in newspapers. Dueling made for good copy. Italian newspapers thrived on stories of duels that the journalists themselves often participated in. When Hauser spoke of the "fever of his profession," it explained in part why Bruneau de Laborie devoted one of the longest chapters in *Les lois du duel* to the question of how to determine if it was the responsibility of the offending journalist or his editor-in-chief to pick up a challenge.[63]

Given the popularity and the laissez-faire attitude of the French press during the belle époque, spoken or written slander was the most obvious cause for wounded honor. Taking a libel case to court was not only a gamble but also insufficient remedy, and so in November 1911, dueling still remained the ultimate arbiter of a conflict regarding slander. In press-intensive Paris there was a constant possibility of your honor being threatened. Taking such a case to court meant that you could be considered a coward and that you relinquished your honor to the whims of strangers. Libel suits were a poor substitute for the duel, as the purpose of the latter was not to force a retraction or to impose punishment, but to reaffirm personal honor and courage. The duel afforded an immediate resolution to a dilemma that otherwise faced an uncertain resolution in the courts.

But let us return to November 23, and the first of the five duels, the one between Chervet and Daudet. Everything about it, from the initial challenge to the formalities that came with it, was reported in *L'Action Française*.[64] It is important to keep in mind that duels were excluded from the French judicial system. Such separation could be maintained only on the basis of very strict informal rules, of which choice of seconds and public notice were crucial elements. It was only by adhering strictly to certain codes of conduct that the duel remained exempt from criminal law and was not simply labeled assault, manslaughter, or murder.

Fortunately, we know even more about this duel because of a brief Gaumont newsreel available on YouTube.[65] It is only two minutes long, and the quality is far from perfect. But we can make out enough of what is happening. And it is not the movements of epées or hands and arms, or the overweight Daudet sparring with the leaner Chervet, that is the most striking feature, but rather the male camaraderie captured once the duel was over, the fraternal bonding over what Gaumont had presented as a duel over "the behavior of Madame Curie." What the footage does not show is that the duel *au premier sang* ended when Daudet suffered a deep gash in his forearm.

The following day, Gustave Téry challenged Pierre Mortier, another *Gil Blas* editor, to a duel because Mortier had criticized Téry's decision to publish the letters. Jeanne Langevin's brother-in-law Henri Bourgeois, the man who was at the center of the stolen letters, was a witness, and the duel ended when Mortier was slightly wounded in the forearm. This duel, too, can be viewed on YouTube, although now the Gaumont newsreel describes the duel as related to the Curie "polemic."[66]

Marie Curie and Jeanne Langevin, the two women directly implicated in the duels, were completely barred from having any actual involvement in them. Only men could receive and award satisfaction. Whether the duel involved a cuckolded husband seeking vengeance on his rival, a brother dueling for his sister's honor, a son protecting his mother, women were an obvious cause but never duelists themselves. In 1883, a former employer of Sarah Bernhardt's, Marie Colombier, published a novel entitled *The Memoirs of Sarah Barnum*. The content was inflammatory enough to cause Bernhardt's son Maurice to charge into Colombier's home and ask her to name a man he could challenge to defend his mother's honor.[67] Max Weber's wife Marianne experienced the slight of the German papers in 1910, when she was indirectly accused of lacking maternal responsibilities. Max Weber came to his wife's defense and stated that he was prepared to defend his wife's honor in a duel, should it come to that.[68]

Curie had no son and no husband. Paul Langevin was her only possible champion. And one morning he appeared at the home of Curie's close friends Marguerite and Emile Borel, "pale and in a buttoned redingote," telling them that he had decided to challenge Téry to a duel. "It's idiotic, but I have to do it." "I have to find witnesses, so that I can make plans," he continued. "Lend me your wife so that she can give me advice and accompany me until this evening." Borel replied: "Done. Take Marguerite. Just get her back to me before dinner." And Marguerite Borel and Paul Langevin set off in a carriage, trying to find the necessary witnesses. It would take time, but he would find his two seconds in the end—one of whom was future prime minister Paul Painlevé—and after sharing a simple meal together Borel went home and Langevin to Gastinne-Renette to try out his pistol.[69]

Duels were an intrinsic element in the theater of urbanity, and like the Great Ferris Wheel, the *vélodrome* was a typical venue. When Téry and Langevin arrived at the Parc des Princes Bicycle Stadium on November 25, at 10:40 a.m., they both carried pistols and had settled on firing one round at twenty-five meters. The following day, *Le Petit Journal* provided its readers with a detailed report of the reason for the duel, letters that proved infidelity "between Langevin, Professor at the College de France and Mme Curie," this time identified only as "the widow of the celebrated inventor of radium." *Le Petit Journal* noted that it was a white duel and that they would explain why shortly. Then followed a dramatic account of how, after all the necessary preparation of the pistols and the terrain and the seconds, and the ultimate "fire,"

> Monsieur Langevin raised his arm partially as if discharging the bullet toward his adversary, but as he fixed his eyes upon the latter, he could conclude that Monsieur Téry, remaining in profile, had not moved and that he continued to hold his weapon pointed to the ground, in the position it had been at the beginning of "Are you ready?"

"A profound silence followed," the newspaper continued, where nobody seemed to know what to do. Judge and seconds deliberated and spoke to the two duelists. The substance of Téry's response, *Le Petit Journal* wrote, was that he was satisfied that Langevin had shown up, but "as for firing at him, no! This is a family man, and in addition, I hold his person in the highest scientific admiration."[70] Finally, one of Téry's seconds fired a pistol in the air, and photographers began pouring into the arena.

In every way possible—intellectually, politically, emotionally— the two men who faced each other that day seem worlds apart. But they shared one crucial experience. Téry and Langevin were both *normaliens*, products of the École normale supérieure, one of the most powerful educational institutions in France. Téry's emotional outpouring confirmed Ute Frevert's contention that it was not "the outcome of the duel which determined whether or not the duelists were men of honor, but the fact that the duel was staged at all."[71] The duel, then, was a fundamentally performative act, understood by contemporary society as being the ultimate arbiter of a conflict involving individual honor. The honor in question rested on protecting a name—in this case a family name as well as a scientific standard—about to be jeopardized by the actions of a woman who not only carried the name herself but had secured its permanent affiliation with French science. Curie could do many things, but she could not mount that defense.

Two more duels remained to be staged, but it is fair to say that Téry and Langevin's tryst represents the emotional and symbolic culmination of that intensive week. And in light of all the turmoil, the accusations, and the duels, the *affaire* suddenly evaporated from the tabloids with remarkable speed. When the dust had settled on the *vélodrome*, it was almost as if the scandal had never happened. Almost, but not quite.

V

The brouhaha of the Langevin affair completely overshadowed the remarkable fact that on November 7, the Royal Swedish Academy of Sciences awarded Marie Curie her second Nobel Prize, this time in chemistry and without making her share the honor with anyone. The Academy had kept close tabs on the events unfolding in Paris, and while they initially remained calm, correspondence between members of the Academy and their French colleagues became increasingly nervous in late November. Even so, around the time of the Langevin-Téry duel, Christopher Aurivillius made plans for the banquet as if nothing had happened.[72]

And then, dated December 1, 1911, she received a letter from one of her staunch supporters in the Royal Swedish Academy of Sciences, Svante Arrhenius, who now had gotten cold feet and urged her not to "accept the prize before the Langevin trial has demonstrated that the accusations made against you are absolutely without foundation." To this she coolly replied: "the action which you advise would appear to be a grave error on my part. In fact, the prize has been awarded for the discovery of Radium and Polonium. I believe that there is no connection between the scientific work and the facts of private life. . . . I cannot accept the idea in principle that the appreciation of the value of scientific work should be influenced by libel and slander concerning private life."[73] It really was a typical Curie answer. She knew full well that science was just as much about persons as things, that the separation of private and public was an illusion, and that it had been so ever since 1903. But she also knew, from recent firsthand experience, that whereas men could transcend and even master the liminal space where private and public overlapped, women could not.

The 1911 duels may seem like a footnote in Curie's life and a speck, if that, in the eye of modern science. They are fascinating performances of masculinity and honor, a solution for slander and libel that some journalists today perhaps regard with a certain nostalgia. But they are also something more. Téry had accused Curie of deploying "deliberate," "methodical," and "scientific" skills in turning Langevin's head, skills *L'Action Française* considered anathema in the private sphere, where their use dehumanized Curie and alienated her from every female virtue. Conversely, the same skills had earned her two Nobel Prizes, a public display of appreciation if ever there was one.

L'Action Française saw in Curie and the Sorbonne a double threat: a threat against both the family and the nation-state. But the dueling male body was not only enacting private grievances, but also acting guardian against the incursions of Science with a Big S. Sangfroid, authority, self-control, the desire for freedom, all were qualities associated with the duel. They were also qualities heralded by nationalists like Daudet and Téry. And they were qualities Pierre Curie had displayed in his heroic quest for radium, qualities associated with science. Embodied in Marie Curie, however, the ability to think and act scientifically was abhorred, not embraced.

The 1911 events showed that upholding the distinction between private and public was as difficult as upholding the distinction between "pure" and "applied" science. We have private letters being stolen and made public by newspaper editors—by proxy representing the public sphere—editors who subsequently engage in dueling as the ultimate public resolution to what is presented as a form of private/public dilemma. And it is difficult not to get the impression that despite all the efforts on the part of all those involved to enforce the boundary between the two, the ultimate result is in fact a constant conflation of what is perceived of as private/public. Curie is curiously present and absent at the same time. Present, because it was her *scientific* (and therefore public) approach to private life (to paraphrase some of the arguments from *L'Action Française*) that was seen as a threat against the status quo. Daudet wanted to uphold the difference between the private and the public, and Curie argued that her private life had nothing to do with her work in the laboratory. In neither case could the distinction be upheld.

Of course, Curie did not listen to Arrhènius but traveled to Stockholm with Irène and on December 11 gave her Nobel lecture. Filled with I's and me's, her speech carefully delineated the extent of her own work as well as that of her late husband and other scientists of the time.[74] Back in ·Paris, Téry's rantings had abated, and the separation judgment between Jeanne and Paul Langevin on December 8—the text Arrhenius had dreaded— did not mention Curie at all. Only a few days after Curie's departure from Stockholm, Eva Ramstedt, who had worked in Curie's laboratory, wrote that she wished that "all the mean people would leave you in peace," so that Curie could concentrate completely on her work and her family.[75] On December 29, she was "dying, condemned to death,"[76] and hospitalized not only for suffering a nervous breakdown, but also for a serious kidney disease. She would remain in the hospital for two months. It was the worst possible start to 1912.

The Gift(s) That Kept on Giving: Circulating Radium and Curie

The passage across the Atlantic would be comfortable, but uneventful. Perhaps that was all for the best. There had been plenty of excitement preceding Marie Curie's departure for the United States on May 4, 1921. A few days before the SS *Olympic* left Cherbourg harbor for New York with its famous passenger onboard, Curie had received a magnificent sendoff at the Opéra. The "national manifestation," organized by *Je Sais Tout* in her honor but "to the glory of French science," celebrated her coming trip to the United States with an impressive list of speakers and the presence of everybody who was anybody in post–World War I Paris. The event even produced considerable proceeds—estimated at around 50,000 francs—for the benefit of the Radium Institute.[1] Self-congratulatory on having initiated the gala, *Je Sais Tout* was convinced that it signaled "the apotheosis of science and French *esprit*." A "touching poem" by Maurice Rostand rounded off the first and more formal part of the celebration. Written to the glory of Madame Curie, it was performed by none other than Sarah Bernhardt.[2]

Ten years had passed since the Académie imbroglio and the

turbulence of the Langevin *affaire*, and Curie had risen from it all like a phoenix from the ashes. Following her dramatic and extensive hospitalization in late 1911, she was bedridden most of spring 1912. During that summer, she brought Irène and Eve along to vacation with Hertha Ayrton, who had offered the sea-loving Curies sanctuary in Devonshire and Cornwall.[3] The two women had met almost a decade earlier, at a time when both lived in a scientific marriage. Now they could add widowhood (William Ayrton died in 1908) and rejection by an academy to their shared experiences. Hertha Ayrton was denied membership in the Royal Society in 1902 on the grounds that married women were not eligible as fellows. On one important point, however, they had made radically different choices: Hertha Ayrton took out several patents. If you type in her name into the European Patent Office database Espacenet, fourteen hits will result.[4] The Married Women's Property Acts of 1870, 1882, and 1893 had given married British women considerably more control over their property than their French counterparts. In contrast to Curie, Ayrton was also politically active in the suffragette movement, and one exception to Curie's rule of staying out of politics came at the request of her British friend. "I am a member of the Association whose leaders are now in prison," Hertha Ayrton wrote to Curie on May 28, 1912, "and I know these leaders personally, and I look on them with reverence as persons of the utmost nobility of mind and greatness of purpose."[5] She wanted Curie to sign a petition for the imprisoned suffragettes. Together with a number of international notables of the time, Curie joined the ranks of those who asked the British government to treat the suffragettes as political prisoners.[6]

When Curie returned to Paris, life went back to normal, until 1914, when World War I erupted, putting her plans for the Radium Institute on indefinite hold. With armistice, she pretty much had to begin from scratch. And she would do so with the assistance of the United States, a country unscathed by the war, epitomizing like no other the benefits to be had from combining pure and applied science. In addition, there was no other country where modernity, innovation, and celebrity were as wholeheart-edly embraced.

Despite failing health and a constant preoccupation with money, her two trips to the United States during the 1920s en-abled her to further consolidate her persona and legacy. Her first visit in 1921 was especially productive, its purpose being to receive a gift from the women of America, who by popular subscription had collected $100,000 for the purchase of one gram of radium, the most valuable material on earth. Compare it to a crowd-sourcing campaign succeeding in raising the equivalent of the 2013 Nobel Prize sum of $1.1 million for a worthy scientist some-where. In 1921, U.S. women scraped together *three times* the amount Einstein received for that year's Nobel Prize in Physics. And it was all for *her*.

In some sense, Marie Curie now reaped the reward of the de-cision made by her and Pierre Curie in 1902. Twenty years after they had ceded radium to others by abstaining from patenting their discovery and the processes of its extraction, the Curies' disinterested action of sharing information and samples was col-lectively reciprocated when the female populace of the United States gave Marie Curie an equally disinterested gift in return. Yet the disinterestedness that was such a significant part of the interchange between these givers of gifts was in fact a multi-layered vortex of symbolic and financial gestures around both radium and Curie. Curie giving to the world and the United States answering for the world with a return gift was something unique. Peripherally related both to a traditional science econ-omy of academies, prizes, and honorary degrees intermingling with state and industry initiatives, this was a public campaign us-ing all the mechanisms of the mass press and celebrity culture to enlist the emotional interest and monetary investment of women across the social spectrum. An informal gesture with significant formal consequences, the gift turned out to be more complicated and more rewarding than Curie could ever have anticipated.

Curie brought Irène and Eve along on the SS *Olympic* to help her cope with the buzz and the media coverage that awaited her in the New World. But it was the fourth passenger accompanying the Curie trio, the New York socialite and editor of the women's magazine the *Delineator*, who made it all happen: Missy Brown Meloney. Sitting in her luxurious *paquebot* cabin writing to her friend Henriette Perrin, Curie described the petite journalist as an "idealist, who seems very disinterested and very sincere." Preparing for a breakneck, six-week-long tour in the United States, Curie dreamt of summers at L'Arcouest, nicknamed "Sorbonne-plage" because of the small group of Sorbonne professors who descended on the village with their families during summers. She told Perrin that she longed for the blue and calm Brittany seashore, a very different Atlantic from the morose and cold one she had almost been taken across.[7]

I

Curie and Meloney first met in the spring of 1920, when the indefatigable journalist prevailed on Stéphane Lauzanne, editor-in-chief of *Le Matin*, to secure a meeting with the famous but notoriously media-shy scientist. Meloney remembered how their discussion quickly turned to the topic of the United States, a country Curie told her guest she had wanted to visit for some time. Nobody knew better than Curie that since the discovery of carnotite ores in Colorado and Utah in 1913, and before substantial findings in Congo would lead the Union minière du Haut-Katanga to basically monopolize the production of radium in the 1930s, the United States was the world's largest producer of radium. Meloney dramatized the conversation with her usual flare:

> "America," she said, "has about fifty grammes of radium. Four of these are in Baltimore, six in Denver, seven in New York." She went on naming the location of every grain.
>
> "And in France?" I asked.
>
> "My laboratory," she replied simply, "has hardly more than a gramme."
>
> "*You* have only a gramme?" I exclaimed.
>
> "I? Oh, I have none," she corrected. "It belongs to my laboratory."

FIGURE 3. Marie Curie and Missy Brown Meloney, USA, 1921. Photographer: Henri Manuel. With permission from the Musée Curie (Association Curie et Joliot-Curie), Paris.

I suggested royalties on her patents. Surely she had protected her right to the processes by which radium is produced. The revenue from such patents should have made her a very rich woman.

Quietly, and without any seeming consciousness of the tremendous renunciation, she said, "There were no patents. We were working in the interests of science. Radium was not to enrich anyone. Radium is an element. It belongs to all people."[8]

The irony of such "tremendous renunciation" depriving Curie of her precious element was not lost on the entrepreneurial New Yorker. Back home, she knew what needed to be done, but Meloney felt too overcome with emotion to approach her idol directly with her ideas.

At least, that is what Henri Pierre Roché—the author of *Jules et Jim*—told Marie Curie in a long letter, writing on Meloney's behalf. According to Roché, Meloney wanted to attract—"(and she can)," he added in parenthesis to underscore that the offer was not without real substance—a gift to Curie's institute in the form of radium and money. And then, he continued, the American editor wanted to publish a history of Curie's life. Roché quickly added that Meloney would work with just as much zeal to ensure that the gift materialized even if Curie decided against writing her autobiography. Quid pro quo or not, Meloney's solution was clearly a two-front proposition. The gift of radium ensured that Curie could continue her research for the greater good of mankind; agreeing to write some sort of autobiography meant securing control of her fame, and Roché urged Curie to grasp the opportunity because "if you do not tell this story yourself, it will be invented one day." If she abdicated such narration to others, he warned, they would "tell anecdotes" and exaggerate the whole "legend." It was a surprisingly irreverent touch that he put "legend" in quotation marks. Reassuring her that she had complete control over the form such a narrative could take, Roché was only too happy to offer his services as an author.[9]

Back home in the United States, Meloney's first step was to prepare an article about Curie for the January 1921 issue of the *Delineator*. Meloney had already contacted Macmillan, Scribner's, Dutton, and Houghton Mifflin on Curie's behalf and intimated that an advance of a thousand dollar and royalties like the 20 percent Theodore Roosevelt received for his autobiography were the ballpark figures Curie should aim for. It was to be the first of many letters giving the impression that publishers were lining up for the privilege of acquiring Curie's story.[10] A few weeks later, when Meloney sent the finished *Delineator* article to Paris for approval, she asked if Curie—provided the fundraising reached the intended target—would consider traveling to the United States to accept her gift. Such a visit would also, Meloney stressed, "stimulate the sale of your book."[11] The answer came back positive: "If you should be successfull [*sic*] in getting Radium for me," Curie promised to do all she could to come to the United States to receive the gift, and she was "collecting facts" for the book she now confirmed she wanted to write.[12]

By mid-December 1920, Meloney told Curie she had been promised enough money to purchase the radium.[13] However, news that the French government was buying up large quantities of radium made several of her potential donors wonder if the fundraising might be superfluous. Maybe Curie did not need help after all, and Meloney asked her outright: "would the gift of a gramme, from American women, for experimental use, be a duplication of the French gift?"[14] Meloney had approached publishers and lobbied for donations. To cancel the whole thing because France had suddenly stepped up to the plate would mean losing face and missing out on an opportunity to show the world what true appreciation and generosity looked like.

One would expect the urgency of the letter to generate a correspondingly urgent reply, but there was almost a month of silence from Paris. Meloney seems to have subscribed to the principle that no news is good news: in the one-way communication that followed, she told Curie that she had received several pledges for contributions; there was not enough money in the bank right now to buy the radium but she expected to be able to do so in April; she had formed a committee of scientists to supplement the committee of women in charge of Curie's visit; the same publishing houses as before remained interested in her book.[15]

Still no response from Curie on the important question if she still needed the radium. If Curie did not like to talk about money, Meloney had no such qualms. The American editor might have left that first rendezvous with a slightly skewed impression of the Polish-French scientist's poverty, and Curie certainly told her as much in one of their first exchanges,[16] but it is hard to understand why Meloney constantly reverted to an economic argument unless she felt it would somehow resonate with her heroine in Paris. Stay away from the lecture bureaus and limit yourself to talks for the science community, she advised Curie, and for two reasons. First, the latter share a common interest with you, and second, and no less important, "they have foundations which enable them to pay you considerable sums for any talks you may be willing to make in their institutions."[17] The two women would not always see eye to eye on things, but in this matter Meloney was a public relation genius, instinctively understanding how important it was for Curie to attract not just any kind of money, but the right kind.

Finally, Curie replied. She thanked Meloney profusely for all her hard work and assured her that the gift, if realized, would be of great use to her. The French radium in question was reserved for hospitals, and it was much easier to procure the expensive element for medical use than it was to keep it for the kind of pure scientific research she had in mind.

Letters during this period reveal a critical juncture in the gifting process, and it is fascinating to see how the gift materializes as a very special kind of commodity, one where giver and recipient openly discussed the exorbitant exchange value of the bequest. Early in their correspondence, Curie had made two things very clear to Meloney. First, it could never be seen as if she had in any way begged for the gift. Asking U.S. women for money would be contrary to the principle of a gift to begin with. Second, Curie wanted Meloney to make sure that the gift in no way cast a shadow on France's commitment to her research. Meloney had no problem complying. But even as a gift to Curie, the radium had to be bought. Somewhere. And as Meloney came to understand from talking to one of the scientists involved in Curie's visit, Dr. Francis Carter Wood, the *where?* had quite a lot to do with the *how much?* He suggested it might be better to simply hand the money over to Curie and let her purchase the radium in Austria, where the price would be lower than in the U.S. market.[18]

But a gift in the form of cash was the one thing Curie did not want. She was adamant that she "would prefer to receive from your Committee the radium that you have purchased yourself, and not the sum destined for its purchase."[19] The exchange of money would have forfeited the purpose of the gift, which was just as much about the recognition of "pure science" as it was about the gifting itself. This mechanism was one Marie Curie understood perfectly. Radium did remarkable things, and if anyone could make it outperform itself even more, it was Marie Curie. Provided she had absolute control of its uses. Exactly what this meant was anybody's guess. After all, a gift is a gift. No questions asked. Besides, only a handful of people worldwide were knowledgeable enough to have any kind of informed opinion on how to best use the radium. The general public would simply have to place their trust in Curie, believing that she would do the right thing. And they did so because she was associated, not with the radium craze of popular culture in the shape of countless books, plays, as well as cosmetics, but with the fluorescent properties and the constant mysterious radiation that captured the imagination from the start. Stories of how Pierre Curie once brought a vial with him to dinner, or how the couple sneaked back into the makeshift laboratory one evening just to sit and watch a tube glow in the dark, contributed as much to the myth of radium as it did to the myth of the Curies. The enormous respect Curie had accumulated since then was the kind of symbolic capital that was incompatible with monetary transactions. To make an explicit link between the two would mean that Science became tainted and lost its disinterested aura.

All of this cultural baggage is why it was essential for Curie to receive *the radium* rather than the money for its purchase. Whether the radium should be bought in the United States or Europe was inconsequential to her. Not to Meloney, however, who was in the less enviable position of having to raise funds to match U.S. prices, even though she knew that Belgium had recently bought Russian radium for $40,000 a gram.[20] Faced with the prospect of paying the U.S. market price of $100,000 a gram, Meloney remained confident that she would end up paying much less.[21]

As their plans begin to take shape, Curie wrote a long letter to Meloney on her plans for the future, telling her about the Fondation Curie, the recent initiative by her and Claudius Regaud to attract private donations destined to support and develop the Radium Institute, all under the patronage of the university. Expanding on what would make her laboratory work and her private life much easier, Curie admitted not having "told you until now about these issues because of discretion." Clearly, she had decided to speak more candidly, underlining that the destination of the gift of radium "that you are making to me must be very precise."

With the gift no longer a secret, Curie informed Meloney that "certain newspapers have announced here that the gift is made to the University of Paris, although you have always said it was made directly to me." She wanted Meloney to clarify her intentions and then continued: "If the gift is made to me, it must be indicated in the text of the donation what my power is to dispose of this gift and within what limits." Exercising control of the radium was crucial to Curie, and she desperately wanted to keep it out of the hands of the university. Again, she emphasized that the gift had to be made to her personally. "We can arrange all of this when I arrive," she added, and in a confident afterthought, "the same goes of course for all other gifts I may receive."[22]

It is quite possible that a letter from University of Paris rector Paul Appell only two weeks earlier had something to do with her insistence on ascertaining whose gift she was actually accepting. Appell had learned from Stéphane Lauzanne—the same *Le Matin* editor who had put Meloney in touch with Curie the year before—that $130,000 dollars or almost 2 million francs had been raised in the United States "to give radium to France."[23] Appell did not forget to mention that in this case Curie's laboratory was a placeholder for France, but Curie might well have felt slightly worried that the university would bypass her when it came to receiving the radium. Meloney did her best to remove any remaining doubt: the radium was *"for your own personal use,"*[24] the gift was to Curie personally and not to the University of Paris, and "the deed will be drawn so that no one can interfere with it."[25]

On March 9, the same day that Curie wrote her long letter to Meloney, the *New York Times* reported that the committee in charge of the campaign preferred it if the $100,000 were made up of "many small contributions rather than of the few large ones that could be secured so much more easily."[26] It was a curious sort of declaration, because why would you choose a more difficult path to secure the $100,000, unless the strategy of relying on the generosity of the rich and famous had failed? Meloney had initially intended to raise the sum by enlisting the support of ten wealthy women from her own circle of friends, each pledging

$10,000, but her fundraising among the New York elite had apparently nosedived.

But Meloney had a knack for turning setback into success. It was true that she had misjudged her friends' willingness to open their checkbooks, but for the campaign as a whole her stroke of bad luck was fortunate indeed. Suddenly, it was not the Park Avenue clique but the flapper, secretary, shopgirl, and factory worker of Middle America who would make the research of one of the world's most famous scientists possible, a scientist-celebrity who was also, or so the story went, a woman and a mother just like them. Strictly speaking, she was nothing like them, of course, and one of Meloney's accomplishments was that she invented "ordinary" Curie, a woman who had longstanding relationships with the ultrarich: Andrew Carnegie had funded her work for years, as well as the Rothschilds.

To get mainstream America to open their purses required greater visibility and a transformation of the tour into a much more public event. One proof of Meloney's clever marketing campaign was that she organized a public bidding for the supplier of Curie's radium. As the tour approached, notices appeared in national newspapers requesting sealed bids for "supplying one gram of radium or any fraction thereof for Mme. Marie Curie," reserving the right to reject any or all bids.[27] Three radium-producing companies were said to have participated, and while the final sum was not disclosed, it was reasonable to expect that the winning bid, that of the Standard Chemical Company, included a discount of around $20,000.[28]

Everything was now ready for Curie's arrival. The *New York Times* reported that the campaign was a resounding success, and contributions continued to pour in to an oversubscribed Marie Curie Radium Fund (MCRF), the balance of which had reached $112,500 even before its beneficiary stepped off the *paquebot* gangway.[29] No wonder, perhaps, that when the small entourage disembarked in New York they were met by what Eve Curie remembered as an "enormous mob."[30] The *New York Times* reported that while Marie Curie looked "visibly ill" from seasickness and wore a "severely plain" dress, despite her age of fifty-three she was "energy personified" as she patiently posed in a deck chair for the twenty-six photographers who fired away with their cameras.[31] Not the most reliable of sources during the period leading up to her visit, the same newspaper had described Curie's "golden and abundant" hair and claimed that she had a Swedish mother. By far their most embarrassing mistake was when happily reporting that Pierre Curie was to accompany his wife on the trip.[32]

But there were only four women in the party that came to New York, and in a famous photograph from that day Marie Curie stands between Irène and Eve, one hand clutching her bag, the other holding her hat. Squinting, she looks slightly amused. Or perhaps it is just the effect of seasickness. In either case, on May 11, 1921, Marie Curie finally arrived in America.

II

Little more than a week later, Marie Curie stood on the White House lawn and accepted her American gift from the hands of President Warren Harding. To be more precise, he gave her an ornate "Certificate for Radioactive Material" and a small key to a box. The actual gram of radium was too dangerous to handle and was later delivered straight to the SS *Olympic* on her departure. In his speech, Harding emphasized that the women of America only gave back to Curie what she had given the world.[33] A return gift, Lewis Hyde explains, "is the final act in the labor of gratitude, and it is also, therefore, the true acceptance of the original gift."[34]

Pictures from that day show what looks like a serene and happy Curie holding her certificate in one hand and Harding by the other arm. Whereas the certificate proved the gift, it was another document, signed the day before all the pomp and circumstance, that was the really significant record of the gift, and it had nothing to do with the president. On May 19, 1921, the Executive Committee of Women of the Marie Curie Radium Fund, represented by Meloney, and Madame Marie Curie of Paris, met in New York and signed the deed stipulating the conditions of the gift in front of the two witnesses Elsie Mead and Grace Coolidge, wife of Calvin Coolidge, the man who succeeded Harding in 1923. Ensuring that "the fullest scientific use be made of such material," the executive committee—acting on behalf of the subscribers—"does hereby give, grant and transfer to *Madame Marie Curie* the said gram of radium to be used and applied by her freely and in her discretion in experimentation and in the best interests of science by herself personally or under her direction."[35] The text could not have been more attuned to Curie's wishes. It was made out to her, not to the university, and it was for "free and untrammeled use" by her in "experimentation and in pursuit of science," giving her precisely the control she had been so eager to secure all along. The text of the deed even found its way into *Science*.[36]

This all-important document would understandably feature in Eve Curie's *Madame Curie*. On the eve of the White House ceremony, Eve Curie described her mother's sudden, dramatic change of heart. Curie now wanted the deed modified, ensuring that the radium offered to her "by America must belong to science." If the deed was left in its current state, it would become "the patrimony of private persons . . . my daughters," when she died. "This is impossible. I want to make it a gift to my laboratory." Curie was about to give away her gift before she had even officially received it and asked a dumbfounded Meloney: "Can we call in a lawyer?" Meloney reassured her that it could all be taken care of next week. Marie Curie's answer has a touch of high drama. "Not next week. Not tomorrow. Tonight. The act of gift will soon be valid, and I may die in a few hours." And, Eve Curie continued, a "man of law, discovered with some difficulty at this late hour, drew up the additional paper with Marie. She signed it at once."[37] Hinting that death might be just around the corner and desperately trying to change the wording at the eleventh hour so that her laboratory and not she personally stood as the recipient of the gift, Marie Curie was making a radical departure from her earlier insistence that the deed be made to her and not the university.

Managing the public perception of how she negotiated her personal and public interests via the gift was of crucial interest to Curie, and to Meloney, who had designed the campaign around a skillful calibration of how this astronomically expensive gram of radium should be used: for experimentation and pure science. Such words implied a freedom from expectations that the American gift wanted to be associated with but still could not fully embrace. It was always intended as a gift to Curie personally, but it was not as if its "free" uses were completely free, after all. Pure science was all very good, but what was expected of Curie was not just any kind of research, any kind of speculative use for the advancement of theory, but research that would eventually produce a cure for cancer. Thus, the gift was intended to be followed by another gift, where Curie was supposed to labor on cancer so that the cycle could continue until she could give back once more, ensuring that the gift did what it was supposed to do, that it "kept on giving."

Finding a cure for cancer was the kind of applied science that made "pure science" understandable. And it led to spectacular headings in the press, some of which came from Meloney herself, who in the April 1921 issue of the *Delineator* left little to the imagination. "That Millions Shall Not Die!" was a not so subtle reference to the fact that radium was indeed seen as the great healer of the scourge of modern life: cancer. Numbers had been circulating in United States in the spring of 1921 showing that cancer-related deaths were on the increase and that cancer now topped tuberculosis as the one major cause of death.[38] Part of the appeal of the radium campaign was no doubt attributable to the simple fact that cancer did not discriminate between heiresses and housewives. It was a classless disease, or so it seemed.

Meloney did exactly as she had been told, telling her readers that Curie had never asked for this help, that the American gift was a return gift for what Curie once gave herself, and as such had been given willingly, gladly, in fact, by the women of America. She used the so-called radium book, where institutions and individuals contributing to the fund signed their names, as an inspired way of promoting her egalitarian rather than elitist campaign. Consider, for instance, the emotional effect packed into her description of how "a wealthy woman walked into the Equitable Trust Company with one thousand dollars to deposit . . . , and her name was listed next that of one of the cleaning women of a down-town office building, who had put down one dollar."[39]

With the popular subscription angle in place, the campaign became a triumphant, truly Capraesque moment in which the general *female* public became a major funding body of science. Inherited or hard-earned, it was the money of American women en masse that would buy a unique woman, Marie Curie, one gram of radium, radium extracted and produced in the United States, but given in expectation of its being used to pursue exactly the kind of "pure" science that once had resulted in its discovery. But that was not all. This collective of women accepted that the radium in question would leave their country in the hands of a woman who would use it for research, not in the United States, *but in France.* It is a measure of Marie Curie's status at the time and Meloney's exceptional marketing savvy that this idea could be sold, even oversold, to American women.

Everything that the deed so carefully set out—Curie's ownership, her power to decide how to best work with such a dangerous but essential resource and to choose whom to work with—all of this distanced her from the women she had to thank for her radium gift. The deed affirmed legal autonomy over an element the power of which only she could harness to the benefit of mankind. Giving Curie this single gram of radium "may advance science to the point where cancer to a very large extent may be eliminated."[40] Nuance was never Meloney's forte, and now she claimed that American scientists predicted an enormous return on the collectively procured and then collectively relinquished radium.

Curie stayed more than a month in the United States following her visit to Washington, D.C. It takes only a cursory glance at her schedule to realize that it would have exhausted even the fittest. She traveled from New York to the Grand Canyon and from the Niagara Falls to Canonsburg, Pennsylvania, the last perhaps a less expected stop on her itinerary. Celebrated at banquets, awarded honorary doctorates at universities and colleges, she listened to songs composed in her honor and speeches that celebrated her achievements. Judging from her animated face in the photographs documenting her visit to the Canonsburg Standard Chemical Company plant where her one-gram radium gift originated, she was more relaxed and at home discussing with the plant managers than toasting at cocktail parties.

In contrast to Albert Einstein, who also toured the country that spring, the press found her difficult. If he romanced U.S. media and journalists, she had a more strained relationship with the corps. Whereas Einstein scattered witticisms around him and played the violin, Curie was tired and withdrawn, canceling talks and visits during the latter part of the tour. The press sarcastically referred to her being unaccustomed to the "strenuous sport" of American "small talk."[41] She was admired, but her sternness, black dresses, and haggard looks did not sit well in a country where celebrity culture depended on celebrities' willingness to play along.

One of 2,031 passengers carried by the SS *Olympic* on the most extensive passenger list departing the United States since before the war, Curie left New York with Irène and Eve on June 25, 1921. Among the 4,200 trunks and smaller packages, and the 1,700 baskets of fruits and candy sent by friends of the passengers as sustenance, one piece of cargo stood out. Broken down into infinitesimal pieces, her radium had been placed in small vials and then locked into a lead box with a casing five centimes thick, now stored in the *paquebot* strong room.[42] In addition to the radium, she had received mesothorium and other ores estimated at $22,000 and cash of $6,684 as remuneration for talks and awards. The insurance letter for her total cargo was for $137,437.[43] It may have been an exhausting tour, but nobody could say it had been a waste of time, financially speaking.

III

So, she came, she was seen, and according to the U.S. press, she underperformed. But then what? Back in Paris again, she had nothing but fond memories for the reporters greeting her at the Saint-Lazare train station, telling them how grateful she was to the women of America for their inestimable gift to science "through me."[44]

Curie definitely returned to France richer in commodities and cash than when she left. In fact, she had even left behind an impressive surplus in the fund set up for the purchase of her radium. As the *New York Times* reported, having raised the sum of $60,000 (somewhere around $731,000 today), and still in anticipation of a promised $50,000 donation from an anonymous but "prominent American," the campaign had been *too* successful.[45]

While there is no record in Meloney and Curie's correspondence of a discussion of what to do with the surplus, except for a statement by Meloney that "I regret very much that I shall not be able to make the committee do all that we wished,"[46] the two women must have broached the subject once or twice during the six-week visit. Keep in mind that it was of Curie's own choosing that the gift came in the shape of radium, and not hard cash. The small fortune sitting there was not reserved for any minerals, nor did it require her to give any more speeches or talks. It was simply *there*, waiting to be used. But how? The executive committee of the Marie Curie Radium Fund had, already while Marie Curie was still in the United States, taken the decision that the surplus funds would be turned into a trust fund in her name.[47]

During the summer of 1921, letters between Curie and Meloney seem to have crossed or gone astray, and it took some time before Meloney informed Curie that the trust fund was being created and that it would give her about $4,000 a year during her lifetime. Knowing that Curie was planning for how her work was to continue after she was gone, Meloney added that the remaining funds would be spent on the education of "two American students in your science." Not such a bad deal, one might think: forging closer contacts with the United States, while ensuring that a younger generation in another country used existing networks to build their own, future ones. But deep down, Meloney suspected perhaps that it was not the kind of news that would go down well in Paris. After all, the excess funds came from the same account that had bought the gram of radium. It had all been collected for the same purpose and given to Curie at her discretion. Why not simply hand over the rest under the same terms? Trying to appease her friend, Meloney promised to chase down a number of Americans Curie had met but who had not yet contributed to the campaign "in the hope that I shall be able to get from them a real free gift for the materialization of your dream and the hopes of Dr. Regaud."[48] It sounds almost as if Meloney and Curie now thought of the tantalizing sum left over in the trust fund as the "real free gift," one that would come to Curie with no strings attached and go directly into the Radium Institute.

No response from Paris. "I can not imagine what has happened to your letters and to mine." By the end of that summer Meloney's growing frustration with the unreliable postal service increased her fears that Curie sulked somewhere, probably in Brittany. In any case, Meloney's letter from August 29, 1921, was completely out of character. Almost gloating, she first told Curie that publishers now felt that her autobiography was "not personal and intimate enough for a book." Only a year previously, she had anticipated a bidding war over Curie's story. Instead, she repeated an earlier offer from the *Delineator* to pay 50,000 francs for three installments of the autobiography. There is every reason to suspect that Curie by this time knew about the Radium Fund dealings. And she did not like what she had found out. Meloney did her best to smooth things over, telling Curie that she had written about the developments well before the news hit the press.[49] Trying for some kind of reaction, Meloney even hinted that perhaps their communication problem could have something to do with Curie's vacationing at L'Arcouest while the mail stacked up at the Radium Institute in her absence.[50]

The extent of Curie's displeasure became obvious in late September, when she contacted Meloney on the practicalities regarding the installments of her autobiography. In handwriting she added, referring to the committee, "I don't quite agree with these decisions and I will write you a separate letter about this matter."[51] Curie had received her American gift and could not, in any shape or form, be seen to act greedily. She had to remain stoically disinterested with respect to the money that she knew could help with her one major priority: to secure the continued existence of the Radium Institute. A private income and funds to cover the education of "women students in chemistry," even if it would allow both American and French students, was not what Curie wanted. Especially not when the decision regarding candidates was out of her hands.[52]

On October 24, 1921, Elsie Mead, the secretary of the MCRF and one of the witnesses to the deed, wrote to Curie because she had learned of her unhappiness with the committee's decision. Patiently, Mead explained how the "phenomenal response, especially through the good offices of university women," and the discount offered by the Standard Chemical Company allowed them to pay not only for Curie's visit, but her daughters' as well. The committee was convinced that a trust fund would be in Curie's best interest, just as the decision to have the remaining income go to scholarships under the auspices of the International Federation of University Women expressed the wishes of the majority of subscribers. We did not, she continued, "ask you to have any part in our decision for we assumed that would put you in a very embarrassing position, for we all know how unselfish you would be and how you would want nothing for yourself."[53] The diplomatic tone showed that Curie had become a prisoner of her own public persona. She had received the gift and should have been grateful for the radium and the scholarships; instead she wanted the money as well, or rather the right to dispose of the surplus as she saw fit. It did not add up, publicity-wise.

Mead's news arrived while the committee was struggling with an unexpected legal snag. As Meloney carefully explained to Curie, the Equitable Trust Company had informed her that it was against the law in the United States to collect money for one purpose and use it for another. Because the campaign had been about raising money for a gram of radium and nothing else, the "letter of the law would compel us to return to the contributors their proportions of the money which was in excess of the amount needed." But how was this even possible, given that the scheme had attracted so many donations of small sums? Well, it wasn't, and Meloney hoped it would prove enough to reach out to those who had contributed ten dollars and above to get their permission to move ahead with the trust fund, this way including the smaller contributions indirectly.[54]

How many subscribers were there? More than a thousand? Frederic R. Coudert, legal counsel for the committee, asked Meloney around the same time as Mead wrote Curie.[55] Meloney responded by compiling a list of contributors of $500 and upward, aggregating some nineteen names, a list J. N. Babcock, vice president of the Equitable Trust Company, acknowledged receipt of on November 17, 1921. He also reminded Meloney that when they last met she had told him that there was a card record of all the contributors, which she had promised to send to them for inspection. He ended his letter with, "I do not think we are in a position to go ahead simply on contributions over $500," proof that Meloney took the easy way out and began by supplying him with the names of only the wealthiest contributors.[56]

Curie was not placated. She told Meloney that she was still lacking funds and that she "would appreciate very much the possibility of using the extra money of the M.C.R.F for the equipment required by my scientific researchers."[57] One reason why she was unwilling to admit defeat is that she had just received a rejection letter from the Rockefeller Foundation, voting not to

support the Radium Institute.[58] Grudgingly, Curie had to acknowledge that while she "would have preferred the free disposition of the money," this would not be possible.[59]

As always, Meloney was ready with a solution. Her hands being tied vis-à-vis the trust money, she immediately responded to
Curie's disappointment by mentioning that there was a $1,000
fund available for the purchase of American-made instruments.[60]
Exactly a month later, Curie sent her first wish list.[61] The instruments Curie wanted would "run a little over $1000," but money
was a minor obstacle now that they were on more friendly terms
again, and besides, "I can take care of that very easily here," Meloney promised.[62] And from now on, Meloney would enlist professors, ambassadors, and friends to carry instruments with them
to Curie when they went to France. "Dr. Moore will take the
galvanometer and the picture and clipping books with him on
May 6th," Meloney wrote to Curie in the spring of 1922.[63] When
her good friend and former *chargé d'affaires* at the Belgian Em-
bassy in Washington, Robert Silvercruys, had to carry the galvanometer instead of Dr. Moore (who was busy taking another instrument, "a very large package"), she "knew that with diplomatic
passports there would be no difficulty or delay or rough handling
of this delicate instrument."[64] There is a nice synchronicity to the
fact that the instruments Curie needed for her scientific work
were transported together with the picture and clipping books,
proof of her celebrity. In fact, this intimate rapport between enabling Curie's research and working with various textual outlets
would now, once the deed problem was out of the way, allow
Meloney to devote more time to the second service she wanted
to perform for Curie: see her life in print.

Meloney knew that she was a skilled manufacturer of stories.
And few audiences worked as hard as Curie's. A group of women
graduates in chemistry at the University of Chicago wrote and
asked if they could name their new sorority Kappa Mu Sigma
after Curie, and if she would agree to become their first honorary member. She accepted.[65] Three typists at the *Delineator*
copied her *Pierre Curie* manuscript, refusing any remuneration

for their work. But they accepted inscribed photographs.[66] There was almost nothing that Meloney could not get people to do for Curie or that they would not volunteer to do themselves on her behalf. Julie des Jardins has argued that the boundless admiration for Curie resulted in a kind of inadequacy backlash. Traces of such borderline anxiety between hope and hopelessness can be seen in a letter Curie once received from Meloney's secretary, Ruth Beard Addis. Thanking her for the inscribed photo she had received, saying it was a constant inspiration to her, Addis continued, "it makes me feel that the very best I can do in this world is not nearly good enough, but that it is worth while to keep on trying."[67]

Curiously enough, Meloney seems first to have learned that Curie had written a biography of her husband after the U.S. tour was over, despite the fact that Curie had signed a contract for *Pierre Curie* with Payot already on September 1, 1920.[68] Meloney immediately wanted to know if Curie would be "willing to let this be published in America? I think it would be a valuable little book."[69] The Payot contract offered a book priced at 3 francs, a print run of 3,000 copies, a 10 percent royalty, and a fixed sum of 1,000 francs, payable in two installments, half on delivery of the manuscript and the other half at the time of publication. The standard clause pertaining to translations was crossed out with a pen in the contract and replaced with a new text in the margin stipulating that Marie Curie reserved all translation rights. The alteration came about toward the end of 1921, when Curie told Payot that she was almost done with the manuscript, and also requested two changes to the contract; to receive 50 free copies instead of the 30 stipulated originally and to reserve all foreign rights, ceding only the French publication to Payot.[70]

Even though Meloney did not know about the biography until a few months later, the translation clause may still be the result of Curie's contacts with Meloney and the increasingly international presence that would follow from an English-language version. But getting the "Autobiographical Notes," and *Pierre Curie* into print would prove another bumpy ride for Meloney in her relationship with Curie. Although she had helped Curie get a contract with Macmillan for her autobiography,[71] Meloney would increasingly nudge Curie to publish it together with *Pierre Curie*. Knowing that Curie's own story ran only 25,000 words, Meloney was worried that it would make a very small book, and perhaps not "personal enough" for the American audience. In addition, she was certain she could get a better contract if the two were published together.[72] Meloney heard nothing from Curie during the summer of 1923 and grew increasingly desperate. There was no news on the autobiography, and definitely no manuscript. With *Pierre Curie* now translated into English, Meloney had managed to secure a better contract from Macmillan combining both texts and garnering Curie $400 more in advance than stipulated in the first offer.[73] Everybody was just waiting to go to press.

In Paris, evidence suggests that Curie struggled with her autobiography and its relation to the *Pierre Curie* book. Even when she admitted that it might be convenient to publish the two in one volume, and the publishers waited for her own story, she seemed reluctant to give the final go-ahead.[74] She knew, of course, that readers were just as much interested in her as in her husband, if not more, and maybe this caused her some sort of writer's block. Whereas she dictated the original *Pierre Curie* to Irène, for "Autobiographical Notes" Marie Curie bypassed any help and wrote directly in English. She may have thought long and hard about how to express herself in a foreign language, but her three typewritten drafts of the "Autobiographical Notes" are not those of an English-language novice. And when she decided to tell the story of the gifting of radium also in her own autobiography, she did so by an almost verbatim repetition of the *Pierre Curie* quote that opens chapter 1 in this book. When the "Autobiographical Notes" returns to the opposition between patenting and not patenting, constructing once again the Curies' well-known ethics, the three drafts are not that different from one another. But finding the right words for the introductory sentence, "My husband, as well as myself, always refused to draw from our discovery any material profit,"[75] proved altogether more difficult. She was always cautious, but when it came to ascribing agency behind the nonproprietary, disinterested stance, she oscillated back and forth. Twenty years after the fact, who had done what? With Meloney breathing down her neck for a manuscript, Curie weighed each word carefully.

In the first draft, she initially situated herself as an active participant in the events, but then changed her mind, erasing the assertive "as well as myself." In subsequent drafts, she toyed around with formulations that gave precedence to Pierre Curie as the instigator of the principle; she merely "followed his view," "shared his feeling," and "followed his plan." She had second thoughts about using "our" when referring to the discovery of radium, opting first for the almost dismissive "that discovery," before reinstating the more proprietary "our discovery" in the second draft. But in the printed version, any traces of assigning herself a role as passive bystander—"shared his feeling, followed his plan"—have disappeared, replaced by a return to the more assertive "as well as myself" and the definitive "our discovery."[76] When she told American readers about her candidacy for the Académie, she was careful to make it sound as if the initiative was not hers. Otherwise, however, it was essential to reclaim rather than renounce agency.

As Meloney had suspected from the very beginning of their friendship, print underwrote Curie's popularity and sustained the willingness to contribute to her work, and the book would help in obtaining instruments for Curie. In July 1923, Meloney even sent Curie a template for how to write such an appeal to Mrs. Nicholas Brady, one of the members of the committee, where "it would be quite appropriate"

> to list all of the things you need in your laboratory. Make the list complete, and enter as nearly as you can the cost of each item. When I say "item," I mean:
>
> Equipment   (so much)
> Platinum   (so much)
> Other materials, or whatever it is that you happen to need (so much)
>
> It would be very simple then for Mrs. Brady to choose anything she wished to do, or to do the magnificent thing and provide all you need.[77]

No doubt Meloney was convinced that *Pierre Curie* would help revive interest in Curie and ensure that money and instruments found their way to Paris. Ridden with problems, production on the book began in earnest during that spring of 1923. Not only would Meloney have to nag and beg Curie for the manuscript and corrected proofs, Curie spotted a number of factual mistakes in Meloney's preface that she believed could be very disagreeable to the University of Paris and that needed to be corrected.[78] One of the reasons why the *Pierre Curie* book turned out to be such a problem was perhaps its unfortunate timing. Curie was preoccupied with the major celebration in Paris of the twenty-fifth anniversary of the discovery of radium, scheduled for December 26. In conjunction with the anniversary, the Chambre des députés had voted to give Curie a national recompense, an annual pension of 40,000 francs fully reversible to Irène and Eve on her death.[79] At the time of Pierre Curie's death twenty years earlier she had also been offered a pension, but refused it, Eve Curie writes, because she felt she was still young enough to work for herself and for her daughters.[80]

The twenty-fifth anniversary was the highlight of her career, the apex of her scientific life. Meloney wanted badly to go to Paris to be part of the celebration, but was forced to write Curie and tell her that the ship "is just about sailing—without me."[81] Late December 1923, she reported that *Pierre Curie* had been well received, with Smith College even using it as a textbook, an initiative she expected other colleges to follow.[82] It was the proverbial calm before the storm. A few days later, Curie sent a very agitated letter to her friend, having discovered several mistakes in the *Pierre Curie* books she recently had received from Macmillan. "What will President Coolidge think of me," she asked Meloney, "knowing that I have dedicated a book to him with fake documents?"[83]

Curie immediately contacted Macmillan, pointing out through her secretary that one of the images in the book showed Jacques Danne instead of Pierre Curie. Nor were the photographs of the Curies' laboratory at the time of the discovery of radium correct. She wanted to know if a correction could be made in the current edition.[84] Macmillan replied that it was impossible to change anything for the first print run and then blamed Meloney for the misunderstandings. Corrections would be made for the second edition, which was about to go to press.[85] When Curie found that the errors remained in the second edition, and that Jacques Danne still stood in for Pierre Curie, she became livid, telling Macmillan she was prepared to make a public statement to the effect that she had no intention of allowing the press to use an inauthentic picture of her husband they had been told was a fake.[86] When her translator, Charlotte Kellogg, asked her if she was pleased with the book, commenting that "the reviews are very cordial," Curie told Kellogg that she would have been happy with it had it not contained three "unauthentic" images that caused "a lot of nuisance."[87]

There was a related aspect of *Pierre Curie* that Curie was irked about as well. Meloney had asked her for signed dedications to include with the book, and Curie had obliged. But somewhere along the line there had been a misunderstanding here too, and Curie had signed an unnecessarily great number of dedications, something that had caused her more problems with her already cataract-afflicted eyes. Meloney tried to make light of the whole thing,[88] but Curie would have none of it. "Any misunderstanding is definitely on your part," she replied, even enclosing a copy of the list she had received from Meloney six months before, saying she would not have gone to all this trouble on her own accord, since she was not in the habit of giving autographs.[89] All in all, the experience with *Pierre Curie* was not a good one.

With the book out of the way, however unsatisfactory to Curie, the trust fund up and running (ditto), and the constant influx of instruments and money arriving in Paris, the relationship between Curie and Meloney had accomplished a great deal in less than three years. Now that both women were back to business as usual, their correspondence tapered off slightly. Maybe they needed a break from each other and the tensions caused by the publication of *Pierre Curie* and the hundred-plus unnecessary dedications Curie had thought Meloney wanted her to sign.

Maybe Meloney was just as exhausted as Curie. After all, it was not easy being an enabler of research on this scale while running a woman's magazine at the same time. Maybe Meloney's health was the reason for the gap in their correspondence during the spring of 1924. Curie was not the only one who was not well, and for a few months in the spring of 1924, there is only silence.

IV

It took an intervention from Eve to break the stalemate that developed after the *Pierre Curie* débâcle. In April 1924, Meloney told the youngest Curie daughter about a serious illness she had just come out of, knowing it was the kind of news that would likely end the impasse. Curie responded that she had known nothing of this, being convinced that Meloney's silence was because she had been too busy to write.[90] And so the two women picked up where they left off. It was almost as if the unnecessary dedications, the wrong photos, and the failure to give Curie control over the surplus funds had never happened. Slowly, years went by. With less pressing matters to attend to, their correspondence settled into a more balanced rhythm. Meloney never wavered in her commitment to Curie, and she displayed constant vigilance when it came to procuring funds or instruments. She was also very protective of Curie's name and reputation.

Toward the end of 1924, Curie received a letter from Morey A. Park of the Morey Flux and Chemical Company in Wilmington, Delaware. Approached by a certain Mr. Jas R. Lake about manufacturing a hair tonic named "Curie Hair Tonic," Morey A. Park contacted Curie to verify if Lake really had what he said he had: Curie's permission to manufacture and market the hair oil in the North and South American countries. Park informed Curie that Lake had promised him that the tonic in question would be in accordance with Curie's "personal formula, which is chiefly bay rum, castor oil and a few other ingredients which we have not as yet been told."[91] Curie wrote back saying that she did not know Mr. Jas R. Lake and that she did not take kindly to having her name used for a commercial product.[92] On the same day, she also enlisted the support of the faithful Dr. Richard B. Moore, chief chemist with the United States Bureau of Mines, asking him if there was anything she could do to prevent such abuse of her name.[93] "Any attempt to use her name would involve serious consequences for the person so doing," Moore very quickly wrote to Park on her behalf.[94] In his reply, a quite wonderful depiction of hucksterism taken to task, Morey A. Park described how he had managed to set up an appointment with the said Mr. Lake, and how the latter had been unable to present any proof of his connection with Curie. Confronted with Curie's letter disavowing any knowledge of him and being told in no uncertain terms that he was making himself liable, Lake promised to make good on Park's suggestion that he write to both Moore and Curie with an apology and an explanation.[95] Moore then sent a copy to Curie of his strongly worded letter to Lake, which he now believed, together with the intervention from Morey Flux, would prevent the charlatan from annoying her in the future.[96]

While Moore concerned himself with Curie's reputation and name, the ever-vigilant Meloney continued to network for Curie. During a choppy passage home from Europe on the SS *Homeric* in November 1924, she made the acquaintance of Owen D. Young, the head of International General Electric Company. She asked for an interview and did as she always did on such occasions: found a way to talk about Curie.[97] Her promotion paid off handsomely, because a few months later and through their mutual lobbying, the two women managed to secure laboratory apparatuses up to the cost of 75,000 francs from the General Electric Company in France.[98] It was the beginning of a long friendship with Young and an equally stable business relationship with General Electric, and as Curie wrote in 1934 when she had received a batch of *gratis* tubes from the company: "It is a pleasure once more to acknowledge the friendly disposition of the General Electric Company towards me and my coworkers."[99]

Meloney also knew that Curie was trying to raise funds for

her second Radium Institute, this time in Warsaw. In the fall of 1927 plans materialized in earnest, and Meloney asked if Curie would consider coming to the United States once again, "if I arranged to do for Poland what I did for you in 1920."[100] Her fundraising would take longer this time, even if the radium price had dropped to almost half of that in 1921. "We are trying to get fifty thousand dollars, which would buy enough radium to start with, won't it?" she asked Curie.[101] And this time, it did. In large part due to the discovery of radium-rich ores in Congo, there was no longer a shortage of the kind that had propelled radium to its unique commodity status in 1921. Curie could now take home another gram of radium, this time intended for Poland. But the relationship with the Belgian supplier, the Union minière du Haut-Katanga, was a sensitive one, and Curie cautioned Meloney to proceed carefully in the acquisitioning process.[102] Owen D. Young received a condescending reply to his request for a discount when he intervened on Curie's behalf; "today," the company wrote back, "we happen to be the principal producer of radium and have to supply 90% of the world's demands." The current price was actually $60,000 per gram, and, they carefully pointed out, since 1923 they had deposited *free of charge* FIVE grams of radium at Curie's institute. Still, the Union minière was willing to let her buy her gram at the bargain price of $50,000.[103] The fact that the Radium Institute already had five grams of radium at their disposal thanks to the Union minière's generosity indicated the extent of Curie's "strategy of accumulation."[104] There was never enough radium, and if the perception of scarcity made it easier to procure what she needed, so be it.

So in the fall of 1929 Curie returned once more to the United States. A new president, Herbert Hoover, received her at the White House, but to considerable less fanfare than eight years previously. This time, there was no public campaign, and the list of contributors to the $50,000 was limited to a wealthy dozen, including Owen D. Young, John D. Rockefeller Jr., and the Daniel and Florence Guggenheim Foundation.[105] Again, Curie stayed with Meloney. Again, there were visits and dinners, the most notable being Light's Golden Jubilee, the fiftieth anniversary of Edison's incandescent lightbulb.

Curie returned home even more indebted to her American friends, and even more positive about a possible third trip, this time perhaps for a longer period. Stanford acting president Robert E. Swain even asked the University of Paris to support Curie's taking a leave of absence during October 1930.[106] However much both Meloney and Curie might have wanted it to happen, always make plans and always canceling them for one reason or other, and however many universities—Caltech, Stanford, Yale, St. Lawrence—that would welcome her, there never came another opportunity for Curie to leave Paris for the United States.

Poor health was one reason. In fact, sickness is a recurring theme during their fourteen years of correspondence, not only their own or that of their children, but also of mutual friends and their children: at one time or other everybody seems to be suffering from an affliction of some kind that forces them to rest and convalesce. However, when Meloney in 1931 insisted that Curie "really should make a will and should do it now," there is a new note of urgency in their letters.[107] It would take some time for Curie to acknowledge that her declining health now made it imperative for her to ensure that on her death the use of the radium gift should be transferred to Irène, "the most qualified person to represent me in this matter, and to follow my views." However, this time the legal snafu was the same that had concerned Georges Gouy: inheritance tax. Curie told Meloney that the gram of radium needed a possessor and she was only its trustee. If it did belong to her, she would have to pay a hefty inheritance tax unless she signed it over to the university.[108] This letter occasioned a reply from Meloney that is worth quoting at some length. There must be no obstacle placed before Irène, she began, and then continued to make a confession.

When I undertook to raise the money to buy that first gram of American radium in 1920–1921, it was with the sole determination to pay tribute to you, and especially to make life a little easier for you and your children. When you came here to receive the radium, you were reluctant to accept it as a personal gift. I tried to dissuade you, but Elsie did not back me up in my contention that the radium was your personal property and that no impulsive disposition should be made of it. Also, I felt certain of one thing—the women who had contributed to that fund desired to make a gift to you. They had no interest in the University Women's Association, nor even in the University of Paris. They had a great sense of gratitude to and pride in you. I shall never forgive those who interfered with my original plan to have the entire trust fund turned over to you. That got beyond my control, but the memorandum which you drew up in which you said you could not accept the radium as a personal gift, and that you would give it for research work—-that memorandum, signed by you and Elsie Mead and Harriet Eager, I kept in my own possession against a day of cooler judgment. I felt sure you were prompted by a shyness, for which I had understanding and sympathy, but I could not approve it. There was no logical person to whom I could deliver such a memorandum, so I locked it up in my personal file. After our long talk that spring afternoon in 1931 when we sat in the park and discussed many plans, I resolved to go over your letters to me and destroy any I thought you would not want me to give to Bill [Meloney's son]. There were a few very kind and friendly letters I should like to have him keep as a souvenir of a precious friendship and one of the most satisfying experiences of my life. I kept that promise with myself and when I destroyed those letters I came upon the memorandum. I was ill. It seemed likely I would not live very long. I took the initiative, my dear friend, and I burned that memorandum. Then I managed to write a note to be delivered to you in the event of my death. I thought you should know I had destroyed that paper. I believed you would forgive me for assuming to decide that matter myself. Sitting here alone, quietly, typing this letter to you, I am even more convinced my action was wise, just and truly honest. I know the spirit in which that money was given, and I am certain it would be a miscarriage of justice to have it go to any one but you and your children.[109]

At the time Curie's plans for her succession were entering a more serious phase, Meloney's confession that she had destroyed the memorandum she had called in a lawyer to change on the eve of the White House ceremony comes out of the blue. How did Curie react to this remarkable confession? Was she silent? Incredulous? Angry? Did she feel cheated? Betrayed? In a handwritten letter sent from L'Arcouest later that August, one she cautioned Meloney to treat as confidential, Curie expressed nothing of the sentiments above. In fact, she admitted having no memory whatsoever of the memorandum in question. "And if you have destroyed it," Curie continued, "I am sure you did for the best." While on the subject of cleaning house, she begged Meloney "to destroy very carefully most of my letters," making an exception for those Meloney wanted her son Bill to keep as personal mementos. "They are part of me and you know how reserved I am in any feelings," Curie wrote.[110]

For more than a decade, the two women had been involved in arranging the surplus funds left over after the women of America gave Curie $100,000 to buy a gram of radium. Throughout this process, from the first letter Curie wrote wanting to make it absolutely clear the gift was to her and not the university in the spring of 1921, to her change of heart on the eve of the White House ceremony and Meloney's decision to first hide and then burn the memorandum in which Curie gifted the radium to science, to Meloney's remarkable confession of the reasons for doing so and Curie's equally remarkable lapse of memory, it suddenly seems that on this their most important common project the two friends never quite understood each other, never really shared the same interpretation of what the gift meant. Curie wanted the radium for herself and for Irène to work with, not necessarily to possess, certainly not to pay inheritance tax on. Meloney wanted it for Curie and her children, not for the university, not for science in general.

The fate of the trust fund remained uncertain for many years, despite Meloney's having enlisted the help of Herbert Hoover, the new president, to convince the committee members to ensure that the trust fund be turned over to Irène.[111] Irène had always been Curie's assistant, and it was increasingly clear that she would shoulder her mother's work in the future. "I found the University Women a difficult group," Meloney told Curie, adding that she had asked Mrs. Herbert Hoover to help her persuade them to include Eve in the bequest.[112] The women Meloney found difficult were trying to secure the funds for the purpose they had wanted from the start, fellowships for women.

On September 19, 1932, Meloney reported that there was to be a meeting of the original Curie Committee for which she was preparing a statement giving Irène control of the radium. And then she told Curie that she had found a first draft of the memorandum she destroyed and that she was sending it to Curie to jog her memory. Now taking stock of their common history and correspondence, Meloney acted on Curie's advice and admitted to having destroyed forty-two letters, but added, "If you wished it I would destroy every line you had ever written."[113]

In a letter marked "personal" from October 28, 1932, Meloney told Curie exactly what she had done to secure the radium for Irène after Curie's death, asking all the members of the committee to sign a resolution to that effect.[114] But in a letter marked "Confidential," from March 10, 1933, Meloney opened by saying, "This note is a confidential communication and I should be glad to have you destroy it after reading it." One reason for her wish was perhaps that the letter, related to her dispatch of the signatures of almost all of the living members of the committee, contained a frank assessment of Mrs. Rockefeller, who did not want to sign the resolution because she felt that it was unnecessary and could cause some sort of precedent, an argument Meloney found "perfectly absurd." And then there was a handwritten addition: "The deed, a photostat of which I have sent to you gives you all legal rights to the Radium."[115] The committee also made it clear that they preferred that Irène "should have control of this gram of radium after the death of her mother."[116] By March 1933, the fate of the Radium Trust Fund had been settled, once and for all. And on March 25, 1934, Curie signed over the radium to the University of Paris.[117]

It is not difficult to understand how Curie benefited from her friendship with Meloney. Radium, honoraria, a trust fund, instruments, donations; there was no end in sight to what was provided from the United States. But what was in it for Meloney? Maybe she was as disinterested as Curie had described her to Henriette Perrin from the SS *Olympic*. Did her reward come from knowing she had helped the world's foremost female scientist to continue her work? There was that, of course. And then there was the matter of being friends with a celebrity. In the spring of 1922, Curie ensured that Meloney received the Légion d'honneur, the honor both Curies had refused. Meloney was thrilled.[118]

>>><<<

The American gift may have been Meloney's brainchild, but it was the women of America who collected and gave the money. It was also her gift, of course, in a kind of distant, impersonal way. But Meloney had actually made another, much more personal gift to Curie, and very aptly, it came in the shape of a book.

Going through their correspondence, I was first puzzled by the many small checks Meloney regularly sent Curie, until I realized they were royalties for a book entitled *The Log Cabin Lady*, published in 1922 by Little, Brown and Company and subtitled "An Anonymous Autobiography." The first mention of a "little check going to you" for this book is from the spring of 1930. It is not much, Meloney added, "but I am sure you will know what to do with it."[119] I admit having some initial difficulty seeing how the story of a poor young woman in the U.S. Northwest leaving her humble beginnings and becoming part of the New York glitterati had anything to do with Curie. As it turned out, the author was none other than Meloney herself, who in the preface told readers that she had met the lady in question at a 1920 Colony Club lunch, where she had been invited to speak about Marie Curie. Struck by living proof of the powerful, transcendent values of American optimism, hard work, and egalitarian principles, she asked the young woman to tell her story and promised her anonymity, and *The Log Cabin Lady* was born. In a letter from September 1930, Meloney explained to Curie that despite "the terrible slump in American business and especially in the book publishing business," the book had earned Curie $73.48 the past year.[120] Between 1930 and 1934, checks kept coming in with small amounts, often given both in francs and in dollars.[121] Curie always accepted them gracefully and sometimes took care to tell Meloney what the money had been used for, even if she once let it slip that she was not quite sure why she had the benefit of it all.[122]

Meloney had almost singlehandedly funded Curie's research for years, and the American gift, its deed, the succession to Irène, the shipping of expensive instruments, the network of rich friends were all thanks to her. But the most intimate gift she ever sent Curie was the royalties from the *Log Cabin Lady*. A tiny drop in the ocean of money Meloney fielded her way, she sent the final check on June 6, 1934, a month before Curie's death. It was for $57.88 from a "little book [that] goes on working for you."[123]

Intellectuals of the World, Unite!
Curie and the League of Nations

Curie never joined the Académie des sciences, and she used the "Autobiographical Notes" to see to it that posterity knew her side of the depressing January 1911 story. The humiliating defeat by Branly became a minor parenthesis in an otherwise stellar career, much ado about nothing. She had been taken hostage by the misdirected ambitions of others, but vindication came to those who waited. And Curie was patient. "Without any solicitation on my behalf," to paraphrase her expression from the "Autobiographical Notes," in 1922 she made academy history anyway as the first female member of the Académie de médecine. Radium's maturation into "Curietherapy," and the American tour in 1921 that had portrayed the element and also Curie herself as part of the fight against cancer, got her through the door. The trenches and battlefields of World War I had opened it. Often reproduced images show Curie driving one of the *petites curies*, the mobile X-ray units deployed for diagnostic purposes at the front and in the teaching sessions she and Irène held in order to train soldiers how to use the machines. Caring for the wounded in battle had always been the kind of heroism women could aspire to without rocking the boat too much. *La radiologie et la*

guerre, published in 1921, told the story in print. It was published
to generally favorable reviews; *La Presse Medicale* thought it "an
excellent work whose only fault is that it is too short."[1] Hertha
Ayrton felt sure that Curie's experiences would appeal to readers
who never looked at an ordinary scientific work; she wanted the
book to be available in "every language and have as world-wide a
circulation as it deserves."[2] This would not be as easy as Ayrton
hoped. Methuen showed interest in the English-language rights,
but decided against publishing the book.[3] Her French publisher,
Félix Alcan, once told Curie that they received many requests for
translations that never materialized because most foreign publishers thought that no money was involved, only the formal-
ity of asking permission.[4] When they received a proposal from
Budapest regarding *La radiologie et la guerre*, Alcan wrote Curie
that they had little hope that their colleague would accept their
price of 1,000 francs.[5]

On June 23, 1931, Marie Curie looked her Academy peers in
the eye and asked them to support a new right, *le droit du savant*.
Not quite the Marie Curie we have met up until now, or so it
would seem. How is it that the woman who went to such great
lengths to defend disinterested science and renounce patenting
in *Pierre Curie* ended up a decade later advocating an *expansion*
of intellectual property law? It would be premature to answer
that question before we consider how the scientific community
during the interwar years came to engage with the international
organization of intellectual property, scientific knowledge, and
intellectual work. Although this story runs on a parallel chrono-
logical track to that of the preceding chapter, the separation is
practical rather than conceptual. As I will try to argue, Curie's
seemingly contradictory stance on intellectual property grew out
of a synthesis of her two experiences of the 1920s, the American
and the international. The ultimate target for her strategies, however, turned out to be France.

And if her discourse was not inconsistent enough in itself, the
context in which it took place was perhaps even more surprising.
Medicine had always had a tenuous relationship with patent law.

In the original 1791 French *loi* drugs could be patented, but with public health issues taking precedence over intellectual property interests, the 1844 *loi* made them nonpatentable subject matter. Remedies could not be patented, and the institution she was a member of had openly opposed any development to that effect. With radium becoming an important component in cancer therapy, her promotion of scientific property was even more improbable. And yet there she was, "paradox of all paradoxes!,"[6] trying to convince the Académie to take a step in just that direction.

I

But let us begin by turning back the clock nine years, to the first week of August 1922, when the League of Nations' newly formed Commission internationale de coopération intellectuelle (CICI) descended on Geneva for its inaugural meeting. Five days of discussion awaited the nine men and two women who represented the best of the best minds Europe—because with the exception of Professors Dilip Banerjee from India and Aloysio de Castro from Brazil, the representation was unapologetically Eurocentric. Curie had received her invitation to join the new organization from Secretary-General Eric Drummond earlier that spring.[7] One of twelve eminent scholars, diplomats, and scientists approached to serve in an advisory capacity to the League's Assembly and Council in all matters relating to intellectual cooperation, Curie belonged to an elite entrusted with the task of guaranteeing that something so incongruous as a Great War never repeated itself.

By now fully recuperated from her first visit to *Dollaria*, Einstein's moniker for the United States at the time of her second trip in 1929,[8] Curie was struggling with *Pierre Curie* and had resigned herself to the inevitable as far as the trust fund was concerned. As so often was the case when wooed and courted, Curie did not accept immediately. Inundated with requests to support this and that cause, she wanted to know more about what she was getting herself into before she said yes.[9] We know little of her political leanings, and what we know is almost by proxy; several of the men closest to her as friends and colleagues leaned to the left. Paul Langevin joined the Communist Party in his sixties, and her son-in-law Frédéric Joliot-Curie was a card-carrying member for many years. But there was nothing remotely controversial or political about the CICI and the League of Nations. It was the kind of disinterested arena that suited Curie best.

Eric Drummond was quick to read her provisory commitment as positive and assured her that it was pretty much up to the committee to define and organize its own work.[10] Einstein, also an invitee but absent from that first meeting because of a research trip to Japan, was even more reluctant to join and had an on-and-off relationship with the CICI until 1924, when he finally began attending its meetings. Curie's coaxing tone in their correspondence during this turbulent period evidences her determination to secure Einstein's presence in the CICI. Having first met in 1909, they were on friendly terms professionally and privately. He had sent her a kind and supportive letter at the time of the Langevin *affaire*, and in 1913 they even vacationed together with their families in the Swiss Alps. In him, Curie saw an ally when it came to science and international collaboration, but he was also the only one whose fame was on a par with her own and who knew exactly what she meant when she asked "if he did not sometimes think it was exasperating not being able to make a gesture without having it take on global significance?"[11] Like no other scientists at the time, both knew just how heavy—and how rewarding—the burden of celebrity could be.

So where was Curie the person and persona in the summer of 1922, when she had returned from the United States and was about to meet her CICI colleagues for the first time in Geneva? Fully formed as an international celebrity, looking forward to the upcoming triumphant twenty-fifth jubilee of the discovery of radium, and a leader of a major research laboratory, she had twelve more years to live and twelve more years to work. She would spend that last decade of her life cultivating two major networks, both of which exerted a profound influence over her scientific strategies and legacy. Her relationship with Meloney was the first. What started out as hero worship on Meloney's part and reserved appreciation on Curie's survived the emotional rollercoaster surrounding Macmillan's publication of *Pierre Curie* and the trust fund problems, turned into a demand-and-supply river of instruments, grants, and personal gifts from sources that never seemed to run dry, and ended up as warm and sincere friendship. Curie's American 1920s made her wiser, not only to the flexed muscles of the United States as an up-and-coming research superpower, but also to the means she could access Over There as a result of her celebrity, the private donations and instruments provided through Meloney's networks and her circle of rich and influential friends.

The second network is the one I am concerned with in this chapter: CICI and the League of Nations. Although my ambition is to show the extent to which Curie's statement on intellectual property in 1931 resulted from her calibration of these two networks, superficially they inhabited separate worlds. Personal idolization, financial reward, and research funding came her way from friends and benefactors in the United States. At the same time, the CICI provided Marie Curie with another kind of platform, one that allowed her to channel her energy into the future international infrastructures for science.

Slowly but surely during altogether ten meetings that week in August, the CICI began to take shape, and in its final report from that first session the committee identified a number of key areas members felt were in need of specific attention. Three of them—bibliography, interuniversity relations, and intellectual property—were important enough to warrant the formation of separate subcommittees. For reasons that I hope will become clear, this chapter largely revolves around two of these topics: bibliography and intellectual property. Work on the first was warranted, because, as the report noted, "the international organization for scientific documentation, particularly bibliography, is essential for all intellectual co-operation," and "the world of science unanimously desires that such an organization be established as soon as possible."[12] It was to be a task led by Marie Curie and Jules Destrée, Belgian ex-minister of arts and sciences.

The second issue, intellectual property, initially had been inconspicuous enough to fall into the quagmire of "Miscellaneous Questions," but it received a noticeable upgrade during the week. In fact, the committee came to the conclusion that "intellectual property, in general, is not sufficiently safeguarded by existing legislation, and scientific property is not safeguarded at all." In light of how modern intellectual property has developed during the past decades, such a statement has a quaint vibe to it. But the CICI did not stop there. "In the matter of scientific discoveries," it suggested, "it should be held that the idea itself is entitled to be safeguarded and not merely the application of the idea."[13] Trying to extend protection so far back in the creative process as to somehow propertize the discovery of a naturally occurring element such as, say, radium was unheard of in existing intellectual property law and would make scientific property the subject of a longstanding controversy.

At first blush, the work of the two subcommittees seems unrelated. By and large, most subsequent research has also treated the issues of bibliography and scientific property as if they had little to do with one another. I will suggest, however, not only that they were closely aligned but that we can follow this larger, structural coupling alongside Curie's personal trajectory during the 1920s. The bibliographic imperative dovetailed perfectly with the search for protection for scientific property. It was the successful accomplishment of the first that made the second possible. The conflict between the principle that there "must be as few obstacles as possible and as many facilities as possible in all matters concerned with exchange of information"[14] with the will to "protect the idea itself" was at this stage conveniently glossed over.

Curie was a dedicated member of the CICI who took her duties seriously. Absent only twice from altogether fifteen plenary meetings, she threw herself into a number of questions and often had something to say about the various proposals that went before the commission. International work and travel seemed to animate, not tire her. Everything she stood for made her perfect for the concerns of the subcommittee on bibliography. Less obvious was that she would also become invested in the efforts toward protecting scientific property. During her twelve-year stint on the CICI, Curie would work toward integrating bibliography and scientific property as complementary, and not conflicting, facets of the scientific endeavor.

II

As it turned out, there already was an organization dealing with many of the same topics that the CICI now gave itself the mission to address, and it was headquartered in Brussels, where Henri La Fontaine (1854–1943) and Paul Otlet (1868–1944) for more than twenty-five years had worked together in the name of pacifism and internationalism. By the time of the first CICI meeting in Geneva, the two collaborators were overseeing an entire portfolio of initiatives that were all prominently "international" in nature: the International Office of Bibliography; the International Archives; the International Library; the International Museum. It was as if the expanding world of information was held together from Brussels. And in a way it was. In 1920, the Belgian government had put a wing of the Palais du Cinquantenaire at their disposal, and La Fontaine and Otlet finally got what they had wanted for a long time; everything under one roof. The Palais Mondial—Mundaneum was a fitting jewel in their international crown.

Utopian is the best word for some of their startlingly modern and innovative work on documentation and information. Perhaps all a bit too utopian for the liking of the League of Nations, even though it had sponsored Otlet and La Fontaine financially as well as morally. And while the Paris peace conference in 1919 decided that the headquarters of the League of Nations should be established in Geneva, Paul Otlet did his best to lobby for a move to Belgium. In effect, what Otlet really wanted was neither Geneva nor Brussels as the site for the League of Nations: he thought the only rational solution was to build a New City, one that was customized for all the new institutions that came with this truly international life.[15] Switzerland was altogether the wrong choice: "few countries are less cosmopolitan," he once quipped.[16] It was not the first or the last time Otlet would promote the idea of a world city. Ten years later, he had changed his mind on Geneva and even considered it as a possible site for his and Le Corbusier's grandiose (but ultimately doomed) vision of a *cité mondiale*. Then, to add insult to injury, Jules Destrée became the Belgian CICI representative and Curie's partner in the sub-committee on bibliography, an assignment tailor-made for Otlet. The Belgian visionaries were sidestepped on almost all counts.

But there was one area where La Fontaine and Otlet exerted crucial influence on the work of the CICI. Among all their internationalist initiatives undertaken during almost fifty years of collaboration, none was more important than *bibliography*. The essence of the universe of information and, most crucially, the antidote to the challenges of modern science, bibliography was also a matter of national pride. In 1895, the Belgian government created the Office international de bibliographie (OIB), "an institution that could shortly become the principal organ for the intellectual life of peoples."[17] Not quite, perhaps, even if La Fontaine and Otlet a few days later hosted the first Conférence internationale de bibliographie in Brussels, the aim of which was to set in motion the project of a universal bibliography.

From a distance of more than a century, one has to admire their ambition of producing not just another run-of-the-mill bibliography, but the ultimate one, "an up-to-date collection of *everything* that has been written."[18] It did not matter where the information came from or in what medium it was produced; the Répertoire bibliographique universel (RBU) was all-seeing and all-encompassing. Otlet initially outlined five main criteria for this bibliography of bibliographies: completeness (containing past, present, but also anticipating future bibliographies); ease of use (classified according to author and topic); accessibility (hence cheap and pedagogical); editable ("errors and omissions are . . . inseparable from a work so considerable as the RBU"); and cumulative (building on already existing bibliographies). In extension, the bibliography was a tool for deposits, statistics, and, most important, for assuring "authors a better legal protection for their intellectual works."[19] From the outset, an international bibliography of this kind was perceived as a support system for intellectual property. But the main impetus for embarking on the hyperbibliography was an understanding of how the massive outpouring of information that followed from the internationalist momentum related to scientific practice. "Passing from what is known to the unknown, helping oneself to the works of all predecessors in order to push the scientific investigation further, avoiding involuntary repetitions and loss of precious time, these are the legitimate concerns of men of science."[20] With the possible exception of the gender bias of its era, what Otlet wrote in 1895 could have been written yesterday.

Not only did bibliographies help scientists do more and better science, but *bibliography* was a science in itself. So it had been for a long time, La Fontaine wrote in the 1895 conference invitation, "only this science has not adopted a common language . . . nor its units of classification like other exact sciences."[21] Fortunately, a solution had now presented itself, and it came in the shape of standardized 12.5 x 7.5 centimeter cards. With the help of La Fontaine's sister Léonie, by 1895 La Fontaine and Otlet had already collected and arranged more than 20,000 of these cards in duplicates, one copy containing information on the author, one on the topic. Arranged in wooden filing cabinets, this was modern information management at its finest, making printed catalogues seem hopelessly outdated.

In addition to their superior flexibility (cards were easily rearranged as more data kept arriving), they could be shared and dispatched on request, annotated and ready for international offices and libraries to use. The fact that their size was the same as the standard postcard format of the Union postale internationale clinched it: information was and had to become even more mobile. However, it was Melvil Dewey's decimal classification system (DDC), which Otlet and La Fontaine used as inspiration for their Classification décimale universelle (CDU), that offered an "almost definitive solution"[22] to the harsh reality of information overload.

For Otlet, Dewey had constructed a "truly international scientific language,"[23] the ultimate proof that bibliography was science. Even if *science* in the context of this chapter must be understood as denoting much more than the natural sciences, in reference to the DDC a narrower definition nonetheless applies. Supermethodological, ever expandable (just as science had proven to be), on standby to anticipate any intellectual development, Dewey's system really was perfect. And the ultimate perfection lay in the fact that it was international. Its ten numerical categories made sense everywhere, and it was hardly a coincidence that on the several occasions Otlet promoted Dewey's ingenious scheme, he illustrated it with the number 5. "Natural Science" (500) preceded "Useful Arts" (600) in Dewey's original scheme, showing that in the hierarchy of knowledge organization, "pure" science took precedence over "applied." It was an additional bonus that the infinitely expandable system was also a professionalization visualizer, demonstrating that there was a glaring need for professionals who had the expertise to fill each numerical subcategory with the requisite content.

"The bibliography is auxiliary to all the sciences,"[24] wrote Otlet and La Fontaine at the time of the Exposition universelle 1900 in Paris. Between June and September of that year there were 127 international conferences in the city, among them the Congrès international de physique, August 6–10, and the Congrès bibliographique international just a few days later, August 16–18. Pierre and Marie Curie presented their findings on radium at the first, and Henri La Fontaine and Paul Otlet their work at the second. Represented by a large part of what then had grown to almost two million cards, their innovation was impressive enough to win the *grand prix*.[25] Delegates to the physics conference spent two days at the exposition, with half a day consecrated to the biggest attraction of the whole event, the moving sidewalk.[26] Maybe the Curies saw the filing cabinets that Otlet and La Fontaine had brought with them on display.

But then war happened, and everything changed. For four years, Europe was crippled by a pandemonium made possible in no small part by science and innovations. Now this chaos could be replaced with progress. All it took was proper organization. Eric Drummond himself had stated that a universal bibliography was "indispensable work . . . a condition of progress for the sciences,—which are necessarily international," but also "for every effort towards the organisation and the grouping of the intellectual forces of the world."[27] And it is with respect to this broader ambition to organize intellectual labor that CICI's work on bibliography and intellectual property made sense ideologically. "Not only have intellectuals been reduced to a condition unworthy of them, they are also the object of an often outrageous exploitation,"[28] wrote La Fontaine and Otlet in 1921. Such a statement tapped into a strong feeling among certain groups during the interwar years that intellectuals—and scientists belonged in this category—had not received the recognition or the reward they merited for their outstanding contributions to society. Bibliography (organizing and documenting scientific work) and scientific property (controlling and rewarding scientific work) were CICI's two-pronged response. Even though the subcommittee on bibliography would sometimes feel that its work was impeded by an unwillingness to cooperate that Marie Curie found frustrating, the necessity for increased international collaboration on bibliography and documentation was largely uncontested. The idea of scientific property, on the other hand, would compensate for such consensus by proving extremely controversial. And as many times before in the history of international intellectual property relations, the inspiration came from France.

III

In the spring of 1920, the French *droit d'auteur* had been revised to include a *droit de suite,* a form of resale right guaranteeing artists a certain percentage of the profits made on the resale of a painting. It was an attempt to compensate for the fact that too many artists—and France could congratulate herself on having plenty of them—lived in abject poverty while creating their masterpieces and then had to sell these for a pittance before they most likely died destitute. Descendants faced similarly depressing prospects, suffering in silence while the paintings in question continued to make money, much money, when sold to collectors or galleries. As might be inferred from the scenario described above, the *droit de suite* was a subset of *droit morale*, that particular aspect of French copyright law safeguarding the inalienable (even perpetual) moral integrity of the author and the work. Just as inalienable, the *droit de suite* had the same cut-off date when it came to legatees; it was in effect until fifty years after the death of the artist. Jules Destrée—who served on all three CICI subcommittees—had initiated work on a Belgian *droit de suite* already in 1913, work that was aborted because of the war. When it ultimately passed into Belgian law in 1921, setting the tax at 2.6 percent of the sale, there was no mistaking the French influence.

Inspiration for the idea of scientific property came from the recognition that artists and scientists shared the same experience: martyrdom. While their commitment to Beauty and Truth almost automatically ensured them a life in poverty, this was a sacrifice they were willing to make if it meant pushing the boundaries of art and knowledge further. After all, the most famous of all living French scientists, Marie Curie, had set the benchmark standard for heroic struggles when she toiled with the tons of pitchblende residue in the rue Lhomond laboratory. And how had society repaid her? Not well enough. Industrialists had piggybacked on the hard work of scientists long enough. The time had come to change the rules of the game. It was, and it would be repeated often, a simple matter of justice. But justice is very seldom a simple matter.

Lucien Klotz, publicist and secretary in the newly funded Comité du droit d'auteur aux artistes, had in 1921 convened a meeting under the auspices of *Le Journal*, trying to come up with a solution to the problem. The meeting ended with two recommendations: that upon the termination of a patent, the inventor would benefit from a *droit de suite*; and that scientists would be the beneficiaries of a similar law, as yet to be formulated. On April 4, 1922, the Chambre des députés discussed a first draft for such a law, made by deputy Joseph Barthélemy. In eleven articles, Barthélemy's proposal was very much a legal hybrid, suggesting a scheme of recompense to scientists on any industrial applications following from their discoveries, setting the duration of scientific property equal to that of the *droit d'auteur* (fifty years), but also recommending modifications within the patent law of 1844.[29] First conceived as a solution to problems faced by artists, the *droit de suite* was quickly converted into a quick fix also for the dilemma of scientists.

The influential Conféderation des travailleurs intellectuels (CTI) then took the matter a step further. CTI organized eleven different professional categories, among them scientists as well as technicians and artists. The glue keeping these disparate groups and interests together was the concept of the intellectual worker, a worker that CTI secretary-general Henri Weindel chastised for having been too slow in organizing properly. It was partly their own fault that intellectuals were now squeezed between two very organized *blocs*; the working classes and capital. Weindel explained that it was not only a certain education that united "the specialists in commerce and industry, administrators, lawyers, men of science, authors and artists, those whom society has to thank for its ideas, esthetics, intellectual and economic methods, law and order, it is the feeling of assuming together the initiative of progress."[30] It was the kind of rhetoric many in the science community found attractive. Paul Langevin, Emile Borel, Paul Appell—three of Curie's closest friends and colleagues—were actively involved in promoting scientific property.

L'Arcouest regular Emile Borel, the man once threatened with demotion for harboring Curie in his Ecole normale apartment when *L'Œuvre* made the letters between Curie and Langevin public in 1911, was also president of CTI.[31] In his preface to the influential 1923 CTI report on scientific property, Borel spoke of general principles such as justice, reward, and international agreements as the only way forward. Roger Dalimier and Louis Gallié, the authors of the CTI proposal, also never really bothered with the possible practical problems involved in the implementation of their ideas. For them, the role of the intellectual worker went beyond self-interest and guaranteed a harmonious social order. Only when all other conciliatory measures had failed would the CTI consider the possibility of collective action: "a strike."[32]

But there were workers and there were workers in science. And as far as radium was concerned, the next few years would bring a major publicity crisis for the element once hailed by *Le Petit Parisien* as the "modern philosopher's stone." Back in 1903, radium had been thought to possess perfect qualities—"mechanical, 'calorique', electrical, physiological"—that promised to cure any affliction, to stand as a conduit to all scientific progress from then on.[33] Twenty years later, the world was about to learn that being too close to the revolutionary discovery could be dangerous, even fatal. In January 1925, two French engineers, Marcel Demalander and Maurice Déménitroux, died after handling radium and thorium X at a private factory. Since they were workplace accidents, their dependents should have been entitled to compensation, but the insurance company refused to pay. When the Paris Cour d'appel after four years decided that their deaths were not a workplace accident but "a professional illness" for which there was no legal remedy, the Syndicat professionel des ingénieurs français and the Alumni Association of the EPCI launched an appeal to collect money for their surviving families: Demalander's old parents and Déménitroux's wife and young child.[34] Curie joined the organizing committee and donated 1,000 francs.[35]

In the summer immediately following Demalander's and Déménitroux's deaths, their fate came up in an exchange between Curie and Meloney regarding recent events in Orange, New Jersey, where William Meloney's physician had been involved in two cases concerning the death of workers in watch factories.[36] Meloney was anxious to know if Curie had any information that might prove valuable to her husband's doctor. Curie told her that

the two French engineers had died because of defective installa-
tions in the factory where they worked but added that she knew
nothing of the conditions in the luminous watch factories. All
the same, she promised to keep Meloney informed of any relevant news from France regarding the health of radium workers.[37]

The industry in question had taken off during World War I,
when luminous watches and instruments became sought-after
practicalities of modern warfare. The "Radium Girls" painted the
watches and dials with radium-fluorescent paint by forming the
brushes with their lips to get them fine-tipped enough for the
small dials. Made terminally ill by their work, five of them initiated a lawsuit against the United States Radium Corporation.
The suit not only changed U.S. workplace safety rules, but also
proved that radium was a killer as much as a healer. As late as
the spring of 1928, workers were dying in New Jersey, and Curie
received a letter from Florence L. Pfalzgraf of the local *Daily
Courier* asking her desperately if she had not "discovered anything which might benefit these women?"[38] There is no copy of
a reply in the Curie archives. When Pittsburgh millionaire Eben
Byers suffered an excruciatingly painful death in 1932 because of
his many years of taking the infamous patent medicine Radithor,
it proved that even the rich and famous succumbed.

Obviously, Radium Girls were not the kind of workers the
CTI had in mind. "Pure science is the mother of all progress, and
that *le savant* should have a right of property in his discovery, just
as the inventor has in his invention, is the most elementary form
of justice."[39] Curie was not as alone in the science of radioactivity
as she once had been; her own daughter Irène was on her way
toward an illustrious career, and many of the women who had
worked in Curie's laboratory had fanned out into new institutional settings and taken their experiences with them into new
networks. There was also Lise Meitner in Germany, Ellen Gle-
disch in Norway, and Eva Ramstedt in Sweden. But the moral
and legal rhetoric still operated in terms of "he," and the intellectual worker was automatically sexed male. Not only was science

a cult of genius and of masculinity, it now resonated with the particular machismo of the progressive rhetoric of class struggle. In order to sell the proposition that the cerebral output of artists and scientists was not the result only of divine inspiration but also of hard work, Curie, Langevin, Borel, Appell, and others had to define themselves as intellectuals rather than scientists. There was strength in numbers, and if authors and scientists banded together they could redress past wrongs. Because the relation between art and science had already been made on the level of individuals as well as collectives, it was not inconceivable that a law developed in one area could be applied in another and that there were features of the *droit d'auteur* that matched scientists' current needs.

>>><<<

The *droit de suite*, Barthélemy's proposal, the CTI report: the ideas and initiatives floating in the air provided background material for CICI's own report on scientific property, written by the Italian senator Francesco Ruffini and dated September 1, 1923. What were its main elements and why would they turn out to be so controversial? The ideological foundation was a form of *droit de suite* for scientists, a royalty on all industrial or commercial exploitation of their discoveries. Even if the war had impoverished the European science community and many countries now struggled to rebuild an adequate infrastructure, the relationship between the intellectual worker—read scientist—and the industrialist was perhaps more nuanced than outright exploitation. Marie Curie's relationship with industry, most notably Armet de Lisle and the factory in Nogent-sur-Marne in 1904, was beneficial to both her and de Lisle, and when the Curies entered into that collaboration, they were accorded "authors' rights."[40] Her refusal to be associated with South Terras Mines in 1914 because such an alliance could tarnish her academic renown and her insistence that Owen D. Young intervene on her behalf with the monopoly supplier Union minière du Haut-Katanga in 1930 and

that he formulate his request for a discount on the Polish gram
of radium as diplomatically as possible show that the relationship
between scientists and industrialists depended on what power
and networks they were able to call upon at the time.

As praiseworthy as was the underlying ambition to protect the
rights of scientists, figuring out concrete means to do so was difficult. Ruffini placed his faith in the law. He argued that while
innovators and artists were both catered to by patents and the
droit d'auteur, there was no legal protection at present for the *savant* "who discovers a truth from which mankind in extension
will thereafter draw the greatest and most durable advantages."[41]
Scientists were unseen by the law. The general objections against
bringing them and their discoveries into the legal spotlight were
the same, Ruffini argued, as those once made in respect to copyright. Like the sciences, art was also cumulative and collectively
constructed. The same thing could be argued about innovation.
And yet, since the Paris Convention for the Protection of Industrial Property of March 23, 1883, and the Berne Convention for the
Protection of Literary and Artistic Works of September 9, 1886,
the international community had managed to work around such
caveats. But how could individual compensation of the kind now
discussed be reconciled with the fact that "the increasing complexity of science effectively makes every great discovery solidary with
previous work without which it would not have been possible"?[42]
Emil Borel's Gordian knot from the CTI report was difficult to
untie. The question was whether it was possible to identify the legal person, the father of the discovery. Ruffini believed it was, and
he would use an unexpected comparison to support his argument.

> The trial of scientific paternity may be of the utmost difficulty, as
> for that matter any trial seeking to establish paternity. But can we in
> good conscience deny the demonstration of paternity to the woman
> who has been with only one man, only because another woman can-
> not tell, among several men, which one engendered her child?[43]

Ruffini made an *explicit* analogy between paternity searches
"in real life," allowed in French law since 1912, and those steps

FIGURE 4. Marie Curie, Albert Einstein, Robert Millikan, and Gilbert Murray at the time of a meeting of the Commission internationale de coopération intellectuelle (CICI) in Geneva, 1925. With permission from the Musée Curie (Association Curie et Joliot-Curie), Paris.

required in order to identify the individual scientist for whom remuneration was due. Real scientific discoveries were like monogamous relationships, as easy to trace as the claims of a virtuous woman who had known only one man. The author truly was a father, someone who wrote texts, discovered new elements or scientific laws, *and* impregnated women.

The CICI remained convinced that scientific property had to go the same way earlier international conventions in intellectual property had gone, the multilateral way. The Paris and Berne Unions and Conventions had consolidated a dual conceptualization of intellectual property and a double administrative apparatus of convention management, journals, and conferences. The fact that the international community had managed to expand these conventions to accommodate new technologies and new rights told the CICI that there was no reason why future revisions could not do the same for the hitherto unknown entity of scientific property, perhaps even at the upcoming Berne revision conference in Rome scheduled for 1928. Ruffini and CICI ended up with a possible new Union pour la protection des droits des auteurs sur leurs découvertes ou inventions scientifiques, relying on the already existing administrative units of Berne and Paris, and a new convention in the shape of a legal hybrid between copyright and patent. Ruffini claimed that the scheme was different from prizes and awards, and distinct from patronage because it offered payment for services rendered, "a more modern, and at the same time more dignified" way of doing things, as he also put it.[44]

Still, it was altogether unclear what sort of scheme the CICI now had before it. Was it a new kind of *moral* right, or was it a new kind of *property* right? The conceptual muddle of subsuming the very French *"droits des savants* over their works or scientific discoveries" under what sounded more like a common law property regime did not help make things any clearer. The Anglo-American preference would always favor a view of property "created by law, rather than recognized by law."[45] Maybe the CICI scheme was a reward system rather than a legal regime, as forthcoming supplementary proposals would suggest. But if it were more along the lines of a redistribution scheme, a form of taxation, would it operate under the parameters of the nation-state or of multilateral treaties such as the Berne Convention? What was, in effect, the relationship between scientific discoveries and national borders? Because radium had been discovered and isolated on French soil, was France the only country liable to underwrite Marie Curie's research if scientific property became a reality? Or should signatories to this new union and convention agree to treat scientists with the reciprocity of Berne? How would disputes be handled, and by whom? Questions were piling up. Nobody knew the answers.

Perhaps a sign of things to come, the first objection originated from within the commission itself. Written by the U.S. delegate John Henry Wigmore, it was appended to Ruffini's proposal and targeted paragraph 3, the scope of the right. The draft convention stipulated that protectable subject matter included "discoveries, that is, exposés and demonstrations of hitherto unknown laws, principles, bodies, agents or properties of living creatures or matter, and innovations, that is, creations of the mind." Even though the law had not always recognized a distinction between discovery and invention, that distinction was fundamental to the modern patent system, and Ruffini himself noted that in 1923 there was not a legal system around that did *not* exclude discovery from protection. Tampering with such a baseline principle was a red flag for the United States, where patent law prohibited patents on the discovery of an abstract principle or a scientific law.[46] And if the principles of copyright should be invoked, opponents claimed that extending the protection for fifty years after the death of the discoverer would force industry to pay the legatees royalties for such a long period that they would go bankrupt.[47]

During the coming years, years spent trying to tweak the untweakable proposal into a sellable proposition, the subcommittee would send out requests for comments and receive suggestions from the international community. Governments replied with lukewarm attention or outright hostility. The critique of scientific property would simply not go away, and it developed along a very familiar axis; the United States and the United Kingdom on one side, and France and her allies on the other. "I am of the opinion that any further discussion of the matter would simply be a waste of time," wrote J. David Thompson, executive secretary of the American National Committee on International Intellectual Cooperation, to Curie in 1928. He had referred the question of scientific property to the National Research Council, and the result was "a very emphatic rejection of the whole project." The United States would not be party to any international convention, nor would it enact any national legislation as far as scientific property was concerned.[48] "Not feasible and of doubtful desirability" was the unanimous verdict from the various divisions of the National Research Council, transmitted by its permanent secretary, Vernon Kellogg.[49] Likewise, the response from the British government was negative from the beginning. The proposal might have the counterproductive and adverse effect of inducing young scientists to focus only on those topics and areas that were financially lucrative; it was doubtful if those who had rendered service to mankind should be compensated financially for this, and, finally, the practical problems would lead only to endless litigation between scientists and industrialists and hamper progress.[50]

The French considered the Anglo-American contingent part-time internationalists, often dull, sometimes crass. The Anglo-American contingent considered the French hopeless philosophers without any sense of reality. Scientific property only widened an already existing rift between the two. It was only a matter of time, however, before France would get an opportunity to show the world just how much more serious she was about international collaboration than anyone else.

IV

International work was hard work, and there was plenty of it, too. CICI's agenda grew exponentially. For each new plenary session, there were new topics and more pressing issues to consider. Finding questions to discuss or people to discuss them was not a problem, finding the money to do so was. And the League of Nations had none. Late in 1923, Marie Curie endorsed a suggestion that the commission be allowed to solicit and receive grants from individuals as well as governments. She proposed a call for funding that would underline the gravity of the situation.[51] Jules Destrée and the CICI chairman, Henri Bergson, were instructed to issue an appeal. France immediately came to the rescue. In a letter dated July 24, 1924, Minister of Education François Albert magnanimously offered Paris as the site for an executive branch and headquarters for the CICI. Despite some rumblings, the offer was accepted.

When *Le Temps* reported from the inauguration of the Institute international de la cooperation intellectuelle (IICI) in January 1926, one can almost sense the effort it took to keep national pride in check behind the veneer of internationalism.[52] Although the offer was the kind of cultural diplomacy gesture the French have always excelled at, the move would not make things easier for the commission. In the next four years, both the CICI and the IICI expanded their activities and faced increasing criticism for inefficiency and lack of focus, both from within and from outside their respective organizations.

With the work on scientific property taking up much of the emotional and administrative time of the CICI, it is easy to lose sight of the more quiet parallel work that had been going on in the subcommittee on bibliography. Curie had certainly taken her work there seriously. She was especially interested in anything that would facilitate a better overview of what went on in laboratories around the world. Curie's own laboratory was internationally staffed, and the United States was not the only country she traveled to during the 1920s. She visited Brazil, Czechoslovakia, Denmark, and Scotland and communicated across the globe both privately and professionally. She knew that science would become only more international in the future and that it was necessary to stay on top of what was happening in laboratories around the world. Quick, reliable, and brief information was the way forward.

If many of the problems of scientific property were easily anticipated, it was perhaps surprising that the subcommittee on bibliography would struggle to get the information it needed from various national professional organizations. Besides turning to official channels, Curie tapped into her personal contacts as well. She told Vernon Kellogg, who was not only permanent secretary of the National Research Council but, together with his wife, Charlotte, also the American translator of *Pierre Curie*, about the ambition to set up an international conference with journal editors, devoted to forming a kind of international agreement on the form of abstracts and their exchange between different countries. She wanted to know if the United States would consider signing such an agreement and send a delegate.[53] She wrote Jean Strohl at the University of Zürich asking for his experiences with abstracts in zoology, and to "Science Abstracts," to learn more about their work.[54] Only too happy to oblige, Strohl informed her about the work of the Concilium Bibliographicum, inspired by Dewey and taking place in close collaboration with Otlet and La Fontaine.

And even if they never crossed paths at the time of the Exposition universelle in Paris in 1900, Curie definitely met Otlet and La Fontaine in March of 1923, when the CICI subcommittee on bibliography convened in Brussels for its second meeting. Otlet was left with a favorable impression of Curie, and noted with satisfaction her interest in the DDC.[55] That summer, Curie reported that the subcommittee had recommended that publishers precede each article with a uniform abstract, one that could then be printed and pasted onto the standardized bibliographic cards and circulated.[56] La Fontaine and Otlet must have approved.

Despite years of butting heads with professional organizations and publishers, and her less than positive experience with the subcommittee's first own bibliographic publication, the prematurely published and much criticized *Index Bibliographicum*, Curie never wavered in her commitment to bibliography. She remained convinced that the need for bibliographic coordination in the CICI was a permanent one.[57] However, in 1926, she was not only a member of the CICI and the subcommittee on bibliography but also a vice-president and member of the permanent committee of the Comité national français de coopération intellectuelle. The list of members included men of letters, scientists, and politicians, many of whom had a longstanding investment in intellectual property, such as Louis Gallié, secretary general of the Confédération internationale des travailleurs intellectuels, lawyer and co-author of the C.T.I 1923 report; Joseph Barthélemy, author of that first proposal for a law on scientific property in 1922, doyen at the Law Faculty at the University of Paris; Emile Borel, member of the Académie des sciences and president of the Confederation française des travailleurs intellectuelles; and, most importantly, Paul Langevin.[58]

Chair of the subcommittee devoted to the *droit du savant* and the recruitment of researchers set up by the French national committee, Langevin was the right man for the job. He had a number of patents in France as well as abroad and even had experience of litigation. One of the most controversial lawsuits he was embroiled in was in relation to a patent registered during World War I. It provoked an outcry from the Service de télégraphie militaire because it made private profit out of a public need. Interestingly enough, Irène and Eve each received 5 percent royalty on one of his patents.[59] Langevin was as careful about protecting his proprietary rights as he was active in hailing the value of disinterested science.

And while scientific property was going nowhere fast in the CICI, support for it in France was still strong. So strong, in fact, that the minister of public education and beaux-arts, Edouard Herriot, in 1928 initiated an interministerial committee whose task was to prepare a draft law on scientific property and the *droit du savant*. Langevin and Curie would both be on the committee, where they would find themselves facing lawyers and industrialists less amenable to such a draft law.[60] Yet it is from this point forward that we can begin to close in on Curie's investment in scientific property, one that in a few years will find an outlet in the discourse before the Académie de médecine. What then was her incentive for taking an interest in scientific property? The most obvious, least interesting, and in the end perhaps most accurate answer is money, pure and simple. But explaining Curie's interest in scientific property as a deplorable means to arrive at an ultimately beneficial end—securing funds for the Radium Institute—tells us nothing of how she managed to reconcile the most significant decision she and Pierre Curie ever made, the decision that *made* the Curie myth, the gifting of radium by abstaining to patent, with the promotion of any kind of proprietary scheme for scientific discoveries. In order to understand how she navigated between these two different poles and how her positions ended up the way they did in 1931, it is necessary to bring Meloney and Curie's U.S. networks back into the picture.

>>><<<

Apart from occasional letters from Curie telling her friend that she is en route to Geneva for CICI meetings, any discussion between Curie and Meloney of the League of Nations or international relations is sporadic and limited to the beginning of the 1930s. Excited about having hired Walter Lippmann to write for her new employer the *Herald Tribune*, in 1932 Meloney was as unbridled in her admiration for the man she called a "philo-sophical journalist" as she had been in her admiration for Curie. She wanted them to meet because "your interest in the peace of the world and war seems to be stirring."[61] About a decade too late in her observation, Meloney thought that it was individuals, not international organizations or agreements that made a dif-ference in the world. Her failure to appreciate fully the extent of Curie's CICI commitment fits with how Jules Destrée felt about the U.S. contribution to the commission as a whole. You would have expected, he wrote in the Belgian *Le Soir* in 1929, that the United States would have helped the commission in some of its undertakings. But no, nothing much had come of that. Once again, one was forced to admit that intellectual cooperation was European. "It has been and will be, *inevitably*. Just as well to be upfront about it."[62]

But the United States had been a good friend to Curie, not only in enabling her research through various contributions in kind and in laboratory equipment, but also in setting in motion Curie's own production of text commodities. And that produc-tion was as essential to the consolidation of her scientific author-ity on the world stage as it was to the circulation of Curie the persona, and because these two were inseparable, Curie would increasingly have to invest in the policing of the Curie name.

Despite her anger at Macmillan for the wrong pictures and credits in *Pierre Curie*, Curie continued to take an active interest in writing for various American publishers. Even when things did not quite work out the way she wanted. Approached by *En-cyclopedia Britannica* to write a contribution entitled "Radium: Its Discovery and Its Possibilities" for what became a two-volume book entitled *These Eventful Years: The Twentieth Century in the Making As Told by Many of Its Makers*, in 1924 she found herself once again quarreling with a publisher. First, Curie refused to sign the contract after she discovered a clause that she interpreted as signing over her translation rights to the publisher.[63] Then, when she received her complimentary copies, she discovered that the illustrations were wrong. In France, her secretary wrote to *Encyclopedia Britannica*, Curie always had final approval on illustrations. That the same practice did not apply in the United States was an unpleasant surprise to her employer.[64] When Hearst's Universal News Service approached her about a contribution, Meloney cautioned that "your articles may not appear exactly as you write them."[65] Under such circumstances, Curie could not see much point in submitting an article.[66]

Curie consulted Meloney partly because the latter knew the U.S. publishing industry like the back of her hand and Curie did not. But Curie was not only asking Meloney's advice on what publishers and outlets she should consider and which ones she should reject. From 1926 on, Curie consulted regularly with Meloney on the best use of "my time and my name."[67] Four years later, the barrage of letters asking her for signed pictures or autographs had escalated to such a level that she felt compelled to ask Meloney, "Could you not, my dear, suggest some arrangement between us, permitting to decide on matters like those? If I start giving out autographs, I shall get in great difficulties."[68] A few days later Curie reminded Meloney to answer the question of "establishing some rules" on the matter.[69]

Actually, Curie's refusal to sign autographs once landed her in a slightly awkward situation. The *New York Times* had sent a request for her signature on the menu of Edison's Light's Golden Jubilee, where she had been seated next to the newspaper's owner, Adolph Ochs. She would not give it. The *New York Times* insisted; surely, she could make an exception for the owner of the newspaper that "without a doubt largely contributed to and stimulated generous gifts from the women of America for the purchase of radium."[70] Through her secretary, Curie once again demurred and said that she had made all her thanks through "Madame Meloney, du *New York Herald*."[71] In one fell swoop she dismissed a request from the newspaper that had published all those untruths about her in 1921 while making sure that her friend's competing paper was recognized as the one authoritative source through which she addressed her American friends.

Meloney's answer to Curie's question about some sort of policy for handling the ever-increasing number of autograph requests sounds very much like what an agent or manager would suggest: "let us take care of things, do what you do best and let us worry about the publicity." Meloney recommended that all queries for autographs and the like should be forwarded to the MCRF committee. And on the topic of the *New York Times*, she felt that while Mr. Ochs's representative "over-stepped himself when he spoke to you as he did," to be on the safe side she enclosed a draft for a letter Curie might want to send to Ochs to explain a bit better why she never signed the menu.[72] It was not Meloney's style to burn any bridges.

Nothing more is said about the matter. But from a collector who wanted Curie to complete his science collection in return for five shillings,[73] to an organization to aid the "Allied Sufferers in the Great War" who wanted a signed photograph for five dollars,[74] the Curie correspondence contains many requests for autographs that display boundless creativity in trying to persuade her to part with a signature on a letter or a photo. But her name was not for sale, nor was she persuaded by any entreaty appealing to her motherly instincts. One letter written on behalf of "a boy who has been struck down with infantile paralysis," asked for her autograph because it would bring "gladness into the life of one which is filled with sorrow." She sent the author of the letter, Charles Eugene Claghorn, no autograph but *Le radium: Le vingt-cinquième anniversaire de la découverte du radium, 1898-1923* (Paris: PUF, 1923), the book published on the occasion of the twenty-fifth anniversary of the discovery of radium. It is unclear if there ever was a boy confined to his bed and if he could read French, but Claghorn was disappointed and repeated his request from someone who "can not use his hands to write." She would not be swayed.[75] She treated requests for interviews in the same manner. Although journalists had described their monosyllabic meetings with Curie as interviews as early as 1911,[76] toward the beginning of the 1930s she was increasingly asked for her opinion on all sorts of issues. But the policy formulated by her secretary

Madame Razet in 1924 that "Madame Curie does not give interviews, not in the sense that the word is generally used," remained unwavering. Razet acknowledged that journalists were received but only if they talked about technical matters or the Radium Institute; Curie never answered personal questions regarding her life, her tastes, or her projects. Furthermore, Razet ended her letter, she did not "give what you call 'messages.'"[77]

In chapter 2, we saw how the Curie name had been a major concern for Curie since the radium standard. Not only was the name a guarantee of scientific excellence in the science community, it had also brought unwelcome notoriety at the time of the failed candidacy for the Académie and the Langevin *affaire* in 1911. Her experiences as an author had proven time and again how essential it was to protect the name from association with unauthorized photos and wrong captions, and the barrage from autograph collectors for whom the name was little more than a tradable commodity had almost required Meloney to step in to manage the flow of demands. But nowhere was the name Curie more important than in medicine, an area exempt from patenting and the branch of science where she was now accepted into the highest echelons.

In his report on scientific property, Ruffini had used pharmaceuticals in order to demonstrate how the law disadvantaged scientists. Here, the exploitation by the industrialist, "first godfather of the invention for its baptism," was again couched in the language of family. It was not necessarily copyright or patents but the trademark that allowed the industrialist to exploit the discovery "for his own profit."[78] In all the work leading up to the CICI report, in all the discussions on rewarding scientists, possibly by some sort of *droit d'auteur*, possibly by revising the patent law from 1844, possibly by formulating a completely new law, the question of the trademark had flown under the radar. But brands were becoming ubiquitous as a guarantee of a particular product's value, a value that was both economic and symbolic at the same time.

In the spring of 1923 Curie engaged in correspondence with

the U.S. Department of Commerce Bureau of Standards. Somehow, she had heard that a certificate from her laboratory was being used in advertising for a pharmaceutical preparation called "Radione." The Bureau of Standards had never heard of the producing company in question but promised to get back to her with any information regarding the improper use of her name in advertising.[79] Having made their inquiries, they assured her that "such misuse of your name has not been at all common in this country."[80] Her answer speaks volumes on how concerned she was that her intervention be seen in the proper light. "If I have been occupied by this question," she told them, "it has only been in the interest of the public."[81]

Around this time, there had been an important development in the relation between scientific discoveries and patenting, one that Curie no doubt knew about and that went in a very different direction from the one she and her husband had once chosen vis-à-vis radium. In 1922, the University of Toronto secured a patent on the discovery of insulin, both the theurapeutic substance and the process of purification. This not only went against established norms within the university to abstain from privatizing essentially publicly funded research, but also departed from the ethics of the medical profession to place as few obstacles as possible in the way of patient care. However, the University of Toronto argued that in this case, patenting was a form of defense, a market regulation in line with the public health interests of doctors and patients.[82] This was patenting in the public interest that Curie mentioned in her letter to the Bureau of Standards. Then in 1925, the nonprofit tech transfer office Wisconsin Alumni Research Foundation (WARF) at the University of Wisconsin-Madison was formed. Thus, the 1920s brought a new phase in the relation between patenting, scientists, and universities.

What the Curies saw as gifting, others simply saw as abandoning. This was exactly what T. Swann Harding accused Curie of when he wrote in 1941 that her "inverted and distorted sense of probity turned radium over to rascals."[83] From Curie's perspective one of the worst of these rascals, one she considered pursuing

legal action against, was the famous brand Tho-Radia. The cosmetics were marketed under the auspices of Alfred Curie, who
had deposited the trademark Tho-Radia for "all pharmaceutical
products, beauty products and perfumerie" in 1932.[84] He was not
related to the Curies, and it was precisely the confusion of the
homonym Curie that was the problem.

Early in 1934, there is a long letter from the attorney J. L. de
Ricqlès in the Curie correspondance. Curiously enough, there is
no addressee on the letterhead, and yet is it clear from its contents
that the person to whom it was written had sought out Ricqlès as
a friend *and* on behalf of Marie Curie. In the Curie archives, the
letter has not been indexed under R (Ricqlès) but under C (NAF
18447), with "Alfred Curie" as the only information provided. The
unknown recipient of the letter, clearly not Curie herself, had
turned to Ricqlès for legal advice on the potential confusion of
the Curie name and the legal remedies that might be available.
Ricqlès summarized the problem in two questions. Could Madame Curie and Pierre Curie's heirs oppose M. Alfred Curie's
use of his name to facilitate the sale of Tho-Radia? Did they have
some means at their disposal to inform the public of this abusive
use? He dispensed with the second question first. Such a public
message risked putting a dangerous weapon in Alfred Curie's
hands and would not get Madame Curie the result she wanted:
"if I understand it correctly, we seek to avoid any polemic." The
first question was the more important one, but also the more
complex. With the possible exception of illoyal competition, one
can use one's personal name freely, and since "Madame Curie
is not *commerçante* and by no means wants to be considered as
one!," this avenue was closed. However, *la première chambre du
tribunal de la Seine* had in a recent judgment—actually involving
the name Pasteur—stated that while the use of one's personal
name to commercial ends was allowed, "one cannot use the name
in such a way that it is susceptible to create prejudicial confusion
to other person's interests, even when these are simply moral."
The last phrase was in red. The conclusion after the four-page-
long letter: a lawsuit was a distinct possibility.[85]

There is no indication that Curie took the matter further, and however much she wanted to defend the name Curie from exploitation by cosmetics companies or from faulty certificates, and however much she worked in the wings to make the draft law on scientific property happen, in public she remained as much a defender of pure science as she had ever been. At the end of 1929, when the journalist Paul Allard of *L'Excelsior* asked for her opinion on the possible commodification of science, her reply was unambiguous. Referring to the work undertaken by the Ministry of Public Education and Beaux-Arts, Curie admitted that they faced a delicate problem. Everybody agreed on the principles, Curie argued. It was a matter of justice. The problem was that the legislation in question had to be international, even global. She continued: "The day when science becomes interested it will cease to exist! It will commit suicide, become annihilated, sterilized." As so many journalists had done before, Allard returned to the Curies' gifting of radium, the result of which, twenty years later, was that the world-famous scientist had to make appeals to the state and to private benefactors for funding. Did she then, Allard asked, regret not having exercised any kind of proprietary right over radium? Well, she answered, "I have been told that I did wrong and if I had acted otherwise the resources from radium would have equipped very useful laboratories, the support of many researchers, the pursuit of new discoveries." And then, after having talked about the decision as if it had been hers all along, she ended by putting Pierre and "we" back into the picture: "the idea of taking out patents didn't even occur to us."[86] In the Curie archives, there is a handwritten comment on the press clipping: "This article was not submitted for the approval of Madame Curie, as she had been promised."[87]

V

Ultimately, Paul Allard's article was a long indignant call for an increase in research funding, and perhaps it is precisely in this context that we must understand Curie's declaration before the

Académie de médecine in the summer of 1931, when, as *rapporteur* of the institution's recently commissioned report on scientific property, she wanted the Académie de médecine to declare itself "favorable to the creation of the *droit du savant* and express the wish that the recognition of such a *droit* should be hastened by the initiative of public powers."[88] Actually, she had used "treated" in the draft version but changed her mind and opted for the more insistent "hastened" in the printed text.[89] The committee whose report she was delivering had convened following a presentation on April 21, 1931, by the same Lucien Klotz who had set the movement for scientific property in motion a decade previously. Ten years later, Klotz reiterated once more the same arguments of justice, reward, and the inseparability of industry and science. The law of scientific property would appeal to industrialists' higher and more elevated sense of the services rendered by science.[90]

What was the reaction from the Académie that summer of 1931? Were they as underwhelmed as everybody else? In the discussion that followed the presentation, Ernest Forneau, at the time director of the *laboratoire de chimie thérapeutique* at the Institut Pasteur, responded to the proposal by saying that he considered *droit de suite* and scientific property "a danger and a utopia." Many of his objections were the same as those that had been raised against Ruffini's proposal, but he also launched into a scorching critique against the current state of French research and higher education, arguing that France lagged behind other nations because of lack of funds and lack of direction.[91]

Of course, there is another, much more pragmatic way of looking at Curie's promotion of scientific property, and that is as leverage to get something else. It is difficult to see how a proposal with so little support managed to stay afloat for so many years, surviving longer in France than anywhere else. Maybe Curie, Langevin, Borel, and other French scientists felt comfortable supporting what they saw as a scheme of remuneration, not propertization. The *droit du savant* might not have seemed contradictory to Curie for the simple reason that she saw it not as a property regime but as redistribution of profits. Similarly, the kind of proposals that were forthcoming under the heading of *droit du savant* criticized the patent system and embraced the *droit d'auteur*. However incomplete and futile, they were focused on rewarding the creative moment, the discovery. They shifted focus from innovation and industry to discovery and creative spark. This would have been acceptable to Curie for all sorts of reasons, one of which was that she entered into the CICI as a scientist but also as an author. But if it was only a scheme of remuneration, one might ask if there was ever any intention of making the *droit du savant* into reality? Maybe it was all intended to force the hand of the state, to push for a new funding scheme for research, a new national plan for supporting science. Curie would not be around to see how her daughter was appointed undersecretary of state for scientific research by the French government in 1936. Nor did she benefit from the Centre national de la recherche scientifique (CNRS), the major French research funding body Irène Joliot-Curie helped launch in 1939, perhaps the result Marie Curie always wanted when she promoted that impossible notion of scientific property.

>>><<<

In July 1930, the three subcommittees on science and bibliography, university relations, and intellectual property were dissolved, their activities continuing in another form.[92] Marie Curie's last major public appearance in an official CICI capacity was as chairman of the 1933 Madrid conference "L'avenir de la culture," organized by the one subcommittee left standing, the one on *arts et lettres*. Curie summarized the far from modest conference agenda in a few main points: Were culture and civilization in a state of crisis? The answer was yes. How did this crisis relate to the dangers of standardization and specialization, the abuse of intellectual effort, and the lack of coordination of available solutions? What was the future of culture and civilization?[93]

Manuel Garcia Morente, the first speaker, set the tone for what was to come. Science was a threat to culture largely because specialization—a direct consequence of developments in science—was a threat to the broader understanding of the world at a time when everything was "anonymous and standardized." Morente frequently referred to the inability of the masses to discern between the good and the bad, their incapacity to understand how science worked, their exclusion from "true culture" because they had been seduced and doped by a culture of mediocrity.[94] Maybe Curie's often cited words on being one of those who believed science had "a great beauty"[95] were intended to balance Morente's pessimism, but the general tone of the Madrid conference was solemn, worried. For more than a decade, Curie had remained steadfastly committed to the principles of the League of Nations and the CICI, but she had never been afraid to speak her mind on the shortcomings of either. "This is a machine," she wrote Einstein in 1929, "that doesn't work the way we would have liked."[96] From the start she had made it clear that she wanted a more limited agenda and leaner and more efficient organization.[97] In 1933, there was not much left of the optimism that in 1919 made Paul Otlet say about the League of Nations that it had the option *not* to construct the world anew on the "model that existed before the great catastrophe."[98] It had been unable to avert another great catastrophe waiting just around the corner. Hitler had become chancellor in Germany earlier that year, and before long there would be another occasion for science to show its darker side to the world.

In the last spring of her life, Curie kept making plans to meet Meloney and to attend various conferences, but on June 22, 1934, Eve Curie wrote a long letter to Doctor Tobé at the Sancellemoz sanatorium in Haute-Savoie, asking if they could accommodate her mother, who needed rest for two or three months. In light of her mother's reduced circumstances ("you know how civil servants are treated"), Eve Curie wanted to know what practical and financial arrangements Sancellemoz and Dr. Tobé could offer them. She stressed the importance of anonymity. Under no circumstances was Marie Curie's visit to reach the press.[99] Dr. Tobé quickly replied that he would be honored to receive her mother, offering a discount from 175 to 100 francs on their most exclusive room.[100] On June 29, 1934, Marie Curie checked into the sanatorium. She died five days later, on the morning of July 4, 1934. Cause of death was "aplastic anemia," brought on by "long-term radiation."[101] Her final words were "La tête tourne." The head spins.[102]

The year 1934 was also an end of some sorts for Paul Otlet and Henri La Fontaine. That year, Otlet published *Traité de documentation: Le livre sur le livre* (Brussels: Van Keerberghen, 1934), his magnum opus collecting fifty years of visionary ideas in one volume, a book today hailed as the first vision of the Internet. Together with their universal bibliography, which totaled 18 million cards when discontinued, Otlet and La Fontaine were also thrown out of the Palais Cinquantenaire in Brussels. For them, as well as for Curie, these first years of the 1930s were the beginning of a slippery slope for their shared internationalist, pacifist, and bibliographical dream. The Mundaneum would remain a sleeping beauty for six decades, until it reopened as a museum/archive in Mons, Belgium, in 1998. In 2012 the Mundaneum entered into partnership with a company that today provides us with the hyperbibliography Otlet and La Fontaine envisioned: Google.[103]

Epilogue

From her earliest years as a celebrity, Curie's life was documented and transmitted to a rapt audience. During more than three decades, there were discoveries, duels, gifts, jubilees, and deaths to report. There were industrialists, scientists, socialites, publishers, and diplomats willing to tell their side of the story. And there were paparazzi and journalists who covered it all in minute detail. But the memory-making processes that made her into a celebrity during her lifetime did not suddenly stop spinning that July morning in Sancellemoz; quite the reverse.

The starting point for the Curie myth as we know it today was September 4, 1937, when the *Saturday Evening Post* began an eight-part serialization of Eve Curie's *Madame Curie*.[1] Arguably the single most important text shaping Curie's legacy, the book won a National Book Award in 1937, was translated into at least twenty languages, and served as the underlying work for the MGM movie with the same name. The Hollywood star MGM first wanted for the leading role was not so very different from Curie when it came to relations with the invasive press.[2] "I want to be left alone": Greta Garbo's famous tagline could have been Marie Curie's. In the end, neither Garbo nor Aldous Huxley, one of several uncredited scriptwriters, were to have anything to do with the Greer Garson/Walter Pidgeon vehicle. *Madame*

Curie is interesting not only because it was the bestseller/biopic that shaped so much of the public perception of Curie over the years, but also because of how it managed to find its way into the science reference books. Used as the basis for entries on Curie in dictionaries and encyclopedias, reviewed favorably in the most established of science journals, Eve Curie's hagiography was a blockade runner across the boundary between science and non-science and treated as a reliable and factual account in both areas.

George Sarton's 1938 *Isis* review is typical of its reception in the science community. It was an excellent book, he wrote, showing that the lives of the Curies "should be read in the same spirit as people read the lives of the saints." Curie once told Meloney that in France, the biography of her husband was intended for students,[3] and Sarton seemed to be of the same opinion. Reviewing both *Pierre Curie* and *Madame Curie* at the same time for *Isis*, Sarton expected his Harvard and Radcliffe students to "read and ruminate the lives of Pierre and Marie Curie; it may awaken in them, if it be there, the love of truth and the love of science."[4] The daughter's book about her mother, the wife's book about her husband, both texts could teach the next generation something about what a life in science should look like. *Madame Curie* did for Curie what she herself could not do; Meloney's professional judgment was spot-on in the summer of 1921 when she accused the "Autobiographical Notes" of not being "personal or intimate enough." Remember how Henri Pierre Roché once admonished Curie that unless she told her story herself, someone else would invent it and probably exaggerate the whole "legend" angle? Well, someone did. And it turned out to be her daughter.

I

On some level, this has been a book about a person who wanted to limit the material later generations would have to work with. Chunks of correspondence are no longer around, and nothing remains of the letters she and Paul Langevin wrote to one another when they shared that tiny apartment, letters kept by Jeanne Langevin and later destroyed by their son Jean.[5]

We still have Marguerite Borel's recollections from *A travers deux siècles* to give some indication of Curie and Langevin's relationship. It is to this book we owe the often-cited image of a Curie in love, her dramatic appearance one evening for dinner at the Borels wearing a white dress with a rose at her waist as opposed to her standard black frock. However polished and fictionalized, *A travers deux siècles* is one of few sources on this period in Curie's life. It is also a quite charming account of what life was like for the French science elite—the Perrins, Borels, and the Joliot-Curie clan—as they spent their summers at L'Arcouest. Borel insisted that her story, published in 1967 under her pen name Camille Marbo—coincidentally the year of the hundredth anniversary of Marie Curie's birth—had been told with great accuracy, because the "infamous libels sleeping in the archives could be exhumed one day by some publicist and made into a pseudo-scandal."[6] With the benefit of hindsight it seems unlikely that the Curie-Langevin romance would have caused a scandal at a time when sexual mores were a bit less Victorian and Paris was bracing herself for the unrest of 1968. Almost fifty years had passed, but to Borel the danger was real enough. Curie and Langevin never ended up together, but in a storybook ending almost too good to be true, their grandchildren did. Hélène Joliot-Curie married Michel Langevin, and in typical French biographical hyperbole, Marbo wrote that the "mixed blood of Paul Langevin and Marie Curie" now ran in the veins of their great-great-grandchildren Françoise and Yves.

The dramatic events in the fall of 1911 had Curie at the center of a private/public dilemma that largely pivoted around the power of the mass press. Curie's ambition to enter the Académie the previous January was still fresh in everyone's memory when the Langevin scandal snowballed at the end of November. On the cusp of World War I, French gender anxieties were at their absolute zenith. Concerned about depopulation and the crisis of masculinity, they could not have found a more suitable target than a woman who demanded a place in both the public and the private sphere. The Curie name was both at its very highest and at its very lowest during this period. It would reach its pinnacle of disinterested value when it became the denomination for the radium standard. It would fall to its very bottom, at least according to *L'Action Française,* when Curie the foreign widow disgraced the name marriage to an illustrious Frenchman had given her. In 2014, when Curie's status as national French monument is unquestioned, it is easy to forget just how often her foreignness was mentioned, and not only by the extreme right. As late as 1932, the popular magazine *Bravo* wanted her to answer a few questions why she—eminent, no question about it, but a foreigner nonetheless—loved France.[7]

The migration of the Langevin *affaire* from the pen to the sword, the ritual of the duel represented a coded performative gendered public act with longstanding traditions, one in which Curie could participate only by proxy. Rather than a footnote in the history of science, this period sheds light on the continuous formation of Curie's persona in the intersection of private and public, honor and reputation, science and tradition.

Emotional display of the kind we see in Gustave Téry's *L'Œuvre* version of the letters between Curie and Langevin—showing a passionate, jealous, and even scheming woman who wanted to share Langevin's scientific life—are rare, very rare. So rare, in fact, that when you come across something in that genre, you take notice. In 1929, when Curie returned home after her second visit to the United States, she wrote a letter to Meloney describing a meeting she had attended at the Ministry of Public Education and Beaux-Arts. She had gone there expecting to participate in a discussion but had found herself being asked for a briefing on the school program in relation to health and efficiency. "I was terrified," she wrote:

> Imagine me speaking without preparation for an hour or so, on a subject I know, but have not made a special study of it. It was a great

trial. I did my best and got a lot of compliments. I wish you had listened to me, because knowing me, you would have been amused by this unexpected exhibition of my talents. I am sorry that some one has not made a speaking film of it. I would then send it to you.[8]

"Imagine me." "Knowing me." And what did Meloney know about this person, who at sixty-two hated to improvise on a topic she was no specialist in, was happy enough about the compliments she got to almost brag about them, who even joked about a "speaking film," documenting the whole thing so that her friend could have watched and watched again "the unexpected exhibition" of her talents; this Curie is a rare bird indeed. Would we have seen something more of the woman who wrote such a "look-how-well-I-did" letter to her friend in 1929 if Curie *had* written her memoirs? Probably not.

An authorized biography had come up in 1931, when Meloney raised the question of whether Curie should set her records in order. Meloney wanted Curie to know that for income derived from an authorized biography to go to her progeny was "quite customary and just."[9] The year before, Curie had been approached by the Agence littéraire internationale (ALI). Following a visit by their secretary to Mme Curie's, ALI expressed an interest in representing her. Literary agents specializing in the memoirs of celebrities, ALI was convinced that hers would be attractive to all the publishers they collaborated with.[10] Curie said no.[11] Stanley Unwin reached out to the Société des gens de lettres that same year, trying to ascertain if Curie was contemplating writing her autobiography. "As she is only sixty-three," he added diplomatically, "she might resent any suggestion that her work is finished."[12] Through the ever-present Madame Razet, Curie politely turned the offer down.[13] When Methuen contacted her and wanted to start negotiations for publication of the autobiography in England, Curie declined. She would consent only to the publication of *Pierre Curie*. She had accepted to write her own story on the insistance of Mrs. Meloney, who "thought it would accomplish some good." These reasons, she continued, did not apply to France or England, "where conditions are different."[14] The United States was an exception, because there research and science played by another set of rules, celebrity rules. Unpleasant as they were, they came with certain undeniable advantages.

Did Curie talk with Eve about one day writing the biography about her life, just as she talked with Irène about how to continue their common scientific work? Was there even a silent agreement between the two sisters, dividing the work on their mother's legacy between them, with Irène taking the "science" side and Eve the "celebrity" side? Pure guesswork on my part, I admit. That the Curie family has been instrumental in making Marie Curie is a less speculative proposition. Irène Joliot-Curie wrote the preface to her mother's collected works, published in Poland on the twentieth anniversary of her death, just as Marie Curie wrote the preface to Pierre Curie's *Œuvres* in 1908. Remaining letters that have entered into the Curie myth are those between husband and wife—very limited since they were seldom apart—and between Curie and her daughters, reinforcing the image of the family as the most stable unit for a woman scientist and perhaps also satisfying a seemingly endless curiosity about how a woman like Curie managed to juggle children and work.

The narrative around the Curie family as one of the greatest families in science and certainly the greatest French family in science cannot be uncoupled from the circulation of the Curie myth. After all, the Curie clan holds the most impressive Nobel track record of all. In 1935, only a year after Marie Curie's death, Irène Joliot-Curie and Frédéric Joliot-Curie received the Nobel Prize in Chemistry for the discovery of artificial radioactivity. Nobel fame also touched Eve, whose husband, H. R. Labouisse, in 1965 accepted the Nobel Peace Prize on behalf of the United Nations Children's Fund (UNICEF). Eerily enough, the careers of the Joliot-Curies and the Curies have striking similarities. Frédéric Joliot's background was as unconventional as Pierre Curie's, and like his deceased father-in-law he became a member of the Académie des sciences only toward the very end of his life, in 1958, the same year he died. Like her mother, Irène Joliot-Curie never made it into the Académie, and like her mother she died young, from leukemia at age fifty-nine, in 1956.

>>><<<

Missy Brown Meloney knew Curie and her two daughters intimately. The correspondence between the scientist and the editor—more than five hundred sheets—is by far the most extensive in the Curie archives. Meloney's influence over Curie's legacy can hardly be overestimated. It was she who did the most to further Curie's celebrity status and lobby for money and instruments for the Fondation Curie, not only by networking with trusts, foundations, and major corporations, but also by encouraging the informal interventions of wives, sisters, and daughters to the men in power, many of whom were received at the Radium Institute. Mrs. Carnegie followed up on the original Carnegie fellowships in 1930 after Meloney had explained how much the depreciation of the franc had cost Curie's laboratory.[15] Meloney reported to Curie that Mrs. Henry Moses, "one of those very rich Jewish philanthropists in this country whose work is always done anonymously," had provided a fund for Curie in her will.[16] The contact with General Electric and the longstanding relationship with Owen D. Young originated with Meloney, who knew how to work the front and back door at the same time.

"You must have all that you want in your laboratory, a little house in the country and a good car to bring you to Paris every day," Meloney wrote Curie around Christmas 1924.[17] Curie got her car, and while the distance from her home on the Quai Béthune of the Île Saint-Louis to the Radium Institute was a fifteen-minute walk, she wrote to Mrs. Edsel Ford saying it was "lovely and comfortable," and that it would be "extremely useful and convenient."[18] Any change in personal circumstances among the Fords, the Moseses, the Bradys, the Owens—investments going sour or funds mysteriously appearing—Meloney somehow knew about and told Curie. When death struck in one of the families, Meloney cabled Curie. Several telegrams from New York report bad news followed by an immediate suggestion to send condolences. Out of friendship and respect, of course, but Meloney knew how important it was that Curie keep showing an interest in the families that were at the heart of Curie's American backing, both professional and private. Meloney really was a kind of manager, as she kept tabs on benefactors and hounded those who had not lived up to their promises. Meloney procured for Curie, fundraised for Curie, managed for Curie, and, significantly, continued to sponsor both Irène's research and Eve's U.S. contacts with the media and art world.

The American gift in 1921 showed that the funding of "pure" science created odd bedfellows and was a stroke of Meloney's marketing genius. But Curie was more than an apprentice to her American friend when it came to understanding just how important the performative dimension of science was. It was Curie who insisted that the gift would take the shape of radium, that it would not come in the form of money. The radium was absolutely necessary just because it was such a symbolic offering, one that represented an exchange of value, of giving and receiving. Radium was a mystery and *had to remain* a mystery. Money was not. Paradoxically, everybody knew just how much the gift was worth. In fact, the exorbitant price was part of the contradictory dichotomies Curie embodied, between gifting and commodification, between pure and applied science, between being a mother and a director of a laboratory. From the very beginning of its existence as a new element, radium was both gifted and commodified along complementary trajectories. Gifting and commodification were practices, however, that needed to be documented and proven in print or through performance. If we take Marie Curie's words in *Pierre Curie* at face value, then it was *because* they gifted radium that the radium industry was born and the flurry of patenting, branding, and commercial uses of radium took off fairly immediately following their discovery.

Curie was active during a thirty-five-year period when almost everything about the science she thought had great beauty changed; her daughter and son-in-law would find themselves working more directly in the framework of "Big Science." After World War II, Irène Joliot-Curie and Frédéric Joliot-Curie would be involved in large-scale industry-science partnerships of the kind that Curie never quite experienced during the "Little Science" of her lifetime. I have always found it quite nice to think of Meloney's description of *The Log Cabin Lady* as "a little book" that "went on working" for Curie as a metaphor also for what Meloney did for Curie. It was Meloney's special brand of "little science," a term not used derogatorily, that helped make the future "big" science of the Curie Institute possible.

The longstanding struggle over the surplus funds, the trust fund, and the deed to the American radium showed just how complex Curie's relationship to radium was, and how aware she was of the need to control it in the future. Indeed, questions of legacy, heritage, and succession have informed the narrative of this book in a number of overlapping ways. From Georges Gouy's letter to Curie at the time of Pierre Curie's death in 1906, to the decade-long controversies over the MCRF deed, to the gift by Curie to the Sorbonne, the control and succession of radium never ceased to preoccupy her. Irène too, thanks to her marriage contract, was in the position—still without the right to vote—to sign over the control of the American radium to her husband during World War II, so that he would continue "the work of my parents and the work we have done in equal collaboration."[19]

On one level, the constant preoccupation with control seems at odds with having ceded radium to begin with. But Curie was always exceedingly careful about and invested in the ownership, control, possession, and use of radium. Some spoke of her relation to radium in terms of giving birth, of mothering, as if a strange kind of umbilical cord conjoined Curie with her element and gave her some sort of birthright that had nothing to do with any kind of intellectual property but that nonetheless made her entitlement irreproachable. Curie, for her part, did little to dispel the circulation of the analogy, because as she had told the women of Vassar College in a speech a few days after her arrival in the United States in 1921, twenty years after having been discovered and isolated, radium was no longer a baby.[20] As could be expected, images of motherhood were commonly used in the U.S. tour. Under her watchful eyes radium had come of age. Under her watchful eyes it would grow to realize its full potential. At the time of the second Nobel, in 1911, *Scientific American* compared fourteen-year-old Irène and seven-year-old Eve to radium and polonium, as the two new elements were "not the only fruits of this ideal marriage, which was blessed by the birth of two children who already give evidence of inheriting the genius of their parents."[21] Eve Curie described André Debierne's discovery of the element actinium in *Madame Curie* as finding a "brother" to radium and polonium.[22]

From marriage contracts to licensing agreements, from proving the paternity of a patent and describing elements as siblings to being complimented as "the real mother of radium,"[23] the idea of family and inheritance featured prominently in the Curie myth. Family similes such as these operate on several levels. First, they were a kind of sense-making mechanism by which the general public could understand the unorthodox collaboration between the Curies. But also, I think, they functioned as a claim-making mechanism following from particular notions of informal entitlement outside formal intellectual property regimes. In their collaboration, Pierre Curie was the innovator of radium, Marie Curie the collaborator and assistant. Family and marriage were the basic organizing unit—scientifically and legally—at the time of the discovery and isolation of radium.

At the time Curie collaborated with her husband, she was not a legal person. Today, we hardly think twice about the fact that women once were *incapable* and that the legal reality of the Code Civil accorded one of the research team absolute autonomy and the other none. Such injustice is just the fossilized remnants of the past. However, ethical challenges regarding legal personhood are hardly extinct. Advances in twenty-first-century science have not exactly made it any easier for us to think about exactly where the limits of legal personhood reside. That we can be granted the autonomy to control external works, discoveries, or inventions is one thing, but how are we to understand the relationship between the *inside* and *outside* of our bodies at a time of harvested tissue and DNA sequencing?

The public and critical success of Rebecca Skloot's bestselling *The Immortal Life of Henrietta Lacks* (New York: Crown, 2010) is an example of how much twenty-first-century readers take an interest in science, intellectual property, and personhood, and how much these categories have changed since Curie's time. The book chronicles the fate of Henrietta Lacks, an African-American woman who succumbed to cervical cancer in 1951, but whose tissue was harvested by the Johns Hopkins Hospital to produce what is known today as HeLa cells. These unique cells have generated not only scientific breakthroughs but substantial revenue, none of which has benefited Lacks's descendants. Skloot's story resonates so profoundly with readers because it taps into our anxieties about autonomy in an age of hyperscience. From the woman Henrietta Lacks to her HeLa cell line, race, science, and intellectual property become understandable through the narrative of the *person* and our wish to explore exactly where that category begins and ends. Laypersons as well as lawyers confront new ethical dilemmas of how to think about and define the boundaries of autonomy, ownership, and property in what is arguably the most personal "property" territory of all. Science, property, and personhood mediate social relations in a number of complex and contradictory ways, some of which come through the law. The right to property may be a keystone right, one where the body (allegedly) does not matter. And yet the body matters absolutely. It did for Curie and it does for us.

The sexing mechanisms of the law in place at the time of the Curies' marriage provide us with some understanding of how and why she developed ownership strategies relational to the law and to the construction of herself as a person and persona. From her first report for SEIN in 1898 to her posthumously published *Radioactivité* (Paris: Hermann, 1935), Curie learned how to frame her authorship and her authority to maximum effect. She was not always successful. In her angry correspondence with Macmillan, Curie wanted to know exactly how many copies of *Pierre Curie* had been printed, and while it was hardly within her power to buy all the copies or see to it that libraries did not include the book among their collections, I have searched in vain for a copy at the Bibliothèque national de France. The American edition and the French edition even had two different authors: the first Marie Curie and the second Mme Curie. And when *Scientific American* published Curie's third and final *Comptes rendus note*, it did so with Marie Curie, not Madame P. Curie, as the author.[24] The American Marie Curie was less in Pierre Curie's shadow than the French.

II

On April 20, 1995, when the Curies were moved from their modest graveyard site at Sceaux and interred at the Panthéon, Marie Curie became the first Great Woman buried among all the Great Men. The ceremony also ensured the Curies, President François Mitterand said in his speech, "sanctuary in our collective memory."[25] All the hoopla and nationalistic pomp seem out of character for the two main protagonists, who were now, whether they liked it or not, made into secular saints of the French Republic. Still, the kind of celebration that took place under the red-white-and-blue banners was not really intended for the approval of the dead, but for the benefit of the living.[26] As posterior counterweight to all the patriotic feelings in the air—French *and* Polish—it is worth recalling Marie Curie's own words from the Madrid conference "L'avenir de la culture" in 1933: "It seems to me, that rather than defending national cultures, we must defend the international culture, because we are all predisposed to feel national things."[27]

It was a nice touch that the man who spoke on behalf of disinterested science that day—Pierre-Gilles de Gennes—was not only a Nobel Prize recipient in physics himself (1991) but even more appropriately the current director of Pierre Curie's alma mater EPCI. The school where Pierre Curie worked for more than two decades and where the Curies discovered and isolated radium is a big, red-brick building that you cannot fail to notice as you walk through the Curie-intensive fifth *arrondissement* with the Curie Institute, Museum, and Archive all within minutes from one another. There is no trace of the famous "wooden shed with a bituminous floor and a glass roof which did not keep the rain out,"[28] where the Curies worked, but the school that provided them with their disinterested institutional setting still stands.

Today, it is known as the École supérieure de physique et de chimie industrielles de la Ville de Paris (ESPCI Paris Tech), having exchanged the *municipale* for the better sounding *supérieure* and added the generic *tech* as the penultimate Americanized identifier of excellence. As a late modern university, ESPCI is, to borrow Steven Shapin's expression, "a mongrel, the result of historical contingencies."[29] One of the more interesting of these contingencies is how the school markets itself by linking publishing with patenting. "The teachers and researchers of the school are constructing the knowledge of tomorrow and publish *1 article daily* in leading international scientific journals; they invent tomorrow's industry by depositing *1 patent per week*" (my italics).[30]

Because ESPCI acknowledges that "constructing the knowledge of tomorrow" is achieved by publishing, and "inventing tomorrow's industry" takes place with the aid of patents, currently ESPCI quite seamlessly brings together the two entities—publishing and patenting—that Curie wanted to separate so badly in *Pierre Curie*. On the other hand, her active involvement in the CICI and her membership in several national commissions looking at a possible French *droit du savant* explain perhaps her declaration in 1927—in direct opposition to what she wrote in her husband's biography—that there was "no conflict between publishing and taking out a patent."[31]

We might applaud Curie's faith in the peaceful coexistence of the two practices but question its practical applicability. Publishing, disclosure, and sharing of knowledge always came first for her; patenting followed. In the mongrel that is the late modern university, scientists may well find the flow going in just the opposite direction: they may find themselves signing nondisclosure agreements because patents, not publications, are the expected outcome of their research funding.

The convergence of the growth of information and the growth of intellectual property takes us back to the CICI's and Curie's work on bibliographies and scientific property. The bibliometrics and scientometrics that today measure and value science output—which ESPCI explicitly quantifies into one article daily and one patent weekly, or seven articles per week as opposed to one patent—were in large part made possible by the kind of framework for scientific documentation that Otlet and La Fontaine developed. "Science is in need of organization,"[32] as Paul Otlet put it in 1921, and bibliography was the answer.

Otlet and La Fontaine's visions and Curie's work of streamlining abstracts feel completely modern and completely relevant to understanding how the two practices of publishing and patenting interact in late modern science. From that perspective, it is something of a mystery why the Mundaneum and Otlet and La Fontaine themselves still remain as unknown as they are. Could it be because the two men were Belgian? It seems like a flippant suggestion but could explain why French scholars, who have no problems accessing their writings, seem unenthusiastic. The French language, on the other hand, could explain why Anglo-American scholars have largely ignored Otlet and La Fontaine. And language is important, as the shift from French to English in the bibliographic universe and the controversy over scientific property in the universe of intellectual property showed.

>>><<<

But what happened to that other CICI concern, scientific property? It is certainly true that it never materialized in the way that Ruffini and all the French proposals envisioned. However, if we by this concept refer to the fragmentary propertization of research data, be it in the shape of genes or archival material, then scientific property is very much a reality. In fact, we could argue that both the bibliographic universe of scientific documentation (journals, databases, abstracts) as well as science itself (genes, tissue, DNA) have been circumscribed by something that is not scientific property but still manages to look a lot like it.

One of the major objections raised against Ruffini's CICI proposal in 1923 was that it would collapse the distinction between discovery and invention. The difference between a naturally occurring phenomenon and an invention was absolute. Radium was a case in point. But conceptual distinctions we think of as rock-solid are more often than not susceptible to negotiations and gradual shifts in technology, trade, and ethics. Starting with the 1980 famous *Diamond v. Chakrabarty* case (447 U.S. 303), which allowed patenting of genetically modified organisms, for twenty-five years we have seen how the frontiers of science move further into the distant horizon. In addition, we now know that knowledge is endlessly exploitable, because despite all the investment that went into defending what was seen as the truth in the 1920s, some claim that the distinction between discovery and invention in patent law and the distinction between idea and expression in copyright both have collapsed.

One domain where science and propertization find themselves in direct conflict with ethics is that of medicine, drugs, and therapeutical development. The Henrietta Lacks case is one example, but there are many more. In 2001, the Curie Institute opposed a patent granted by the European Patent Office to Myriad Genetics for the so-called "breast-cancer gene" BRCA1.[33] The patent was revoked in 2004. Countries like South America, Brazil, and India have been involved in trade disputes with the United States because they have allowed local companies to produce generic copies of patented drugs.

In the 1920s, scientific property was presented as a question about individuals, even when interested parties recognized the cumulative and collective nature of science. Recompense schemes were geared toward persons, usually identified as the single (male) discoverer. Today, however, the patenting and publishing body is no longer primarily an individual, but a collective. Universities, groups, corporations are the person(s) from which the authority of knowledge and innovation stems. Articles can have hundreds of authors, and patenting has become so complex that it requires the involvement a whole panoply of legal and administrative inputs.

The mongrel university is one of the entities within whose walls investments seem to have shifted from disinterestedness to interest, a development often attributed to the U.S. Bayh-Dole Act in 1980, allowing federally funded universities to patent. But as I tried to show in chapter 4, there had been much earlier examples in the case of the University of Toronto patenting insulin and the WARF tech transfer office in the 1920s. Nobody bats an eyelid when it comes to the patenting ambitions articulated by universities like the ESPCI; in fact, it has become a way for higher education and research institutions to market themselves in an environment where branding strategies are part and parcel of knowledge production.

III

As I tried to argue in the first chapter, it is not as if the links and overlaps between patenting and publishing were not present in Curie's own life, however much she wanted to convince us otherwise. In fact, it is at the precise time Curie put the disinterested legacy into print in *Pierre Curie* that she began to take an active interest in scientific property. Her celebrity status was already cemented; during the 1920s the press wanted her opinion on everything under the sun. If a question like "Does marriage interefere with a career?"[34] was not too implausible, the request from the *Gazette Apicole Montfavet-Avignon* asking if she would contribute to the Christmas special issue celebrating the bee is among the oddest in the Curie archives. Philosophers, lawyers, and humorists had all agreed to write their salute to the hardworking honey maker, but Curie declined to offer the scientist's perspective.[35]

Maybe all of the requests were a sign that Curie had become truly famous, or maybe it was just that celebrity had become truly commonplace. In the introduction we saw Curie chosen the "Most Inspirational Female Scientist of All Time," in L'Oréal's 2009 poll. In 1931 she received "a beautiful traveling case" completely equipped with "Evening in Paris" toilet accessories from another cosmetics company, Parfumerie Bourjois in New York. She immediately cabled and asked for no publicity. Picture her, if you will, this famous scientist and two-time Nobel receipient, receiving a letter promising "Evening in Paris" toilet accessories. The most remarkable thing about the "Bourjois Prize of the Week for outstanding achievement in the woman's world," however, was that it was a *weekly* honor.[36] Celebrity had become ubiquitious and ordinary, that is, truly modern.

Curie's relationship with intellectual property was more complex than she made it out to be in *Pierre Curie*: renouncing patents but making a substantial income from patents while Pierre Curie was alive; embracing copyright and author's rights as a suitable inspiration for the *droit du savant*, and very carefully policing the Curie name. While the demand for autographs evidenced the increasing circulation of Curie as commodity, there were other indicators that 'Curie' was nearing appropriation on a new scale. When she received a request from an R. Cortesi for permission to name a future dispensary in the nineteenth *arrondissement* "Pierre-Curie," Madame Razet answered on her behalf that she never allowed her husband's name or her own to be used for private businesses.[37] On the other hand, she was quite willing to accept it when schools, dormitories, and streets were named after her and Pierre Curie. In 1925 there was a request for her authorization to name Pavillion VII (a dormitory with 58 rooms) of the Cité Universitaire Curie. Her reply was affirmative. A follow-up letter arrived a few days later. Through the intervention of Paul Appell, the Cité Universitaire had been made aware of the fact that it might be appropriate if the authorization covered *both* Curies, not only the husband. Again, she consented.[38] Four years later, the municipality of Trignac wanted to name its new school for girls "Curie." Whereas the boy's school would carry the name "Jean Jaurès," there was no first name attached to the Curie surname. As with the radium standard, the name could refer to either Curie. Pleased to accept, she sent her best wishes for the inauguration.[39] Dispensaries come and go, but schools and streets remain. All commercial uses are rejected, all streets, schools, and institutional requests accepted.

"Brand" is a term that we generally associate with the intellectual property regime of trademarks, and especially in an extended and contemporary form where it has become a placeholder for the multifarious personality, aura, or style associated with a particular product. Using such a designation for a historical person rather than for a product of the universal recognition of, say, Coca-Cola, Nike, or Apple might seem misplaced, but it makes sense, I believe, when seen as an integral part of Curie's intellectual properties. Albert Einstein is the best example of a celebrity scientist making the transition into commodity culture as a brand. This commodity form, which sees "persons and things both become thinglike,"[40] has—despite the occasional t-shirt— not reached the same level with Curie as it has with Einstein.

While Otlet and La Fontaine have not yet become brands, Melvil Dewey has. The Online Computer Library Center (OCLC), provider of WorldCat, another contemporary bibliographical tool, owns "Dewey." The OCLC has trademarked not only "Dewey" but also "DDC" and "Dewey Decimal Classification System." In 1993, OCLC initiated a lawsuit against a small boutique hotel near the main branch of the New York Public Library: the Library Hotel. The Library Hotel organizes its rooms according to the Dewey Decimal Classification System and furnishes them with books according to the category in question. Fifth floor: Math and Science. Room 500.001: Mathematics. And so on. Because the Library Hotel had not acknowledged the OCLC as the rightful owner of the marks in their advertising, OCLC wanted the hotel to pay damages and to acknowledge their proprietorship of Dewey.[41] The settlement fairly quickly arrived at out of court stipulated that the Library Hotel would "receive permission from OCLC to use the Dewey Decimal Classification® trademarks in its hotel and in its marketing materials, with an acknowledgment that OCLC is the owner of the Dewey® trademarks."[42]

Avoiding confusion is the common denominator between this case and the seemingly unrelated Curie–Tho-Radia problem in 1934. The public/customers needed to understand which Curie/Dewey they could trust, and which one they should distrust. J. L. de Ricqlès's letter underscored that Alfred Curie's abuse had to be countered not in the context of commodification but on moral grounds of reputation. Few have described the confusion of the advertising age as well as Jessica Litman, who lists how customers need to be "shielded from confusion about the source of a product at the point of sale, . . . protected from after-market confusion, reverse confusion, subliminal confusion, confusion about the possibility of sponsorship or acquiescence, and even confusion about what confusion the law makes actionable."[43]

Interestingly enough, the one place where Curie today does circulate as a brand is in the European Union. The Marie Skłodowska-Curie Actions is one of the more prestigious of research funding schemes of the Horizon 2020 program. When the program launched as the Marie Curie Actions on Facebook in February 2013, the European Commission explained that this was a strategy to "get across with young researchers and students and further spread out the Marie Curie brand."[44] Not the Marie Curie Actions program brand, but the Marie Curie brand: the disinterested science that the program wants to promote is framed completely in property language. When you take into consideration that the MSCA uses an image of Curie that clearly has been influenced by Andy Warhol's palette and depiction of female celebrities like Marilyn Monroe, Jacqueline Kennedy, and Ingrid Bergman, no one has made a brand out of Marie Curie better than the European Union.

There is no Curie® in the way that there is a Dewey®. Yet the fact remains that both persons/personas, names we associate with the circulation of information, knowledge, and disinterested science, are becoming more commodity-like by the day. The use of Dewey's name by an organization like OCLC and the branding

of Marie Curie by the European Commission show just how so-
phisticated public/private partnerships in knowledge production
have become.

>>><<<

Meloney once told Curie, "You are a world person."[45] Of course,
the American editor idolized her friend, always placing her above
diplomatic entanglements, as untouchable, uncorrupt, and "pure"
as the science she represented. But the Ivory Tower is wrong
for Curie, a woman who consciously and creatively managed her
persona so that it got her what she wanted. Some of her strategies were channeled into print, from scientific articles to popular
biography, while some found an outlet in the international collaboration on bibliography and scientific property.

When François Mitterand reiterated the Curies' many virtues
on that stand in front of the Panthéon in April 1993—zeal, enthusiasm, precision—he chose disinterestedness as the one trait
that described the Curies' scientific ethos better than any other.
Disinterestedness sounds so very grand, and interest so very
bad. But as I have tried to show in this book, the disinterested/
interested matrix in science does not come in white-or-black, but
in shades of gray.

The continuous circulation of Marie Curie the person/persona
rests on a complex interplay between divergent economies, legali-
ties, and values that remains only partially acknowledged. From
the 1903 Nobel to the 1933 Madrid conference on the future of
culture, Curie straddled the border between science and non-
science and between formal and informal modes of control and
ownership. In the last decade of her life, she was not only a fully
formed international celebrity, but also someone who had the capacity to set in motion networks across three intersecting arenas,
U.S., international, and French.

Still, there is a bit too much myth and not enough *mensch*
when it comes to Curie, and Meloney is not the only one who
helped put her on a pedestal. To recognize that we have made
substantial investments in keeping Curie there also means rec-

ognizing that we have been less willing to understand her as an institution builder, a networker of the highest order, the kind of modern scientist that built alliances, attracted other scientists around her, and protected her investments. To embrace the image of a more complex Curie does not make her any less a role model, or her achievements any less impressive. Instead, it could help assign her the same irreducibility, complexity, and wholeness we accord the male scientist before we pick him apart and put him back together again, knowing deep down that he will remain somehow intact through the whole process. So will she. And coming to terms with that Marie Curie could also mean making a modest contribution to rewriting the history of modern science.

Acknowledgments

For the privilege of spending three engrossing years (2010–13) devoted to Marie Curie I owe thanks to HERA (The Humanities in the European Research Area), which allowed me to be part of the research consortium CULTIVATE (yes, another acronym, short for "Copyrighting Creativity: Creative Values, Cultural Heritage Institutions and Systems of Intellectual Property"). Working with project leader Helle Porsdam and research assistant Sigurdur Olafsson at the University of Copenhagen was a true pleasure. Likewise, Fiona Macmillan, Lucky Belder, Madeleine de Cock Buning, and Valdimar Hafstein were all wonderful fellow project investigators. At the risk of sounding unprofessional, I confess that one of the fondest memories from our collaboration together stems from the time when we wrote up the project proposal at Helle's house in Munich. Suffice it to say that it involved food and drink at a Biergarten followed by a walk home through a very Black Bavarian Forest. Helle, Fiona, and Lucky: that's what I call perfect conditions for proposal writing! Another great thing about CULTIVATE was getting to know Mia Rendix, Roeland de Bruin, Aki G. Karlsson, and Stina Teilmann-Lock, who all contributed with their own projects and general good cheer.

Presentations at various seminars, conferences, and workshops helped shape the book into its final form. In Stockholm and Uppsala, I am grateful to Göran Bolin, Katarina Nordquist, Saara Taalas (the event I'm thinking of was in Åbo, but still), Staffan Bergwik, Anders Lundin, and Otto Sibum for their input and assistance. Once again, the Department of Archival Science, Library and Information Science, and Museology at Uppsala University provided the necessary infrastructure and support for my research.

In 2010, Andrew Kenyon extended an invitation to speak at the University of Melbourne, and it was thanks to him, Megan Richardson, Ramon Lobato, Julian Thomas, James Meese, and Simone Murray that my stay in that city became so pleasurable. Matthew Rimmer deserves a gold star for always being in my corner and for inviting me to Canberra to give a talk at the Australian National University during that visit. Special thanks to David Philip Miller at the University of New South Wales in Sydney, who was on another continent at the time I was down under, but who still arranged so that I could present an early version of the project to his history-of-science colleagues. A year earlier, a similar opportunity at the University of Wisconsin-Madison allowed me the double pleasure of reconnecting with Christine Pawley as well as meeting Florence Hsia for the first time. As I was finalizing the manuscript in Paris during the summer of 2013, I bumped into Florence in the reading room of the BnF Richelieu. Serendipitous meeting = great dinner!

Thanks to Janice Radway I could kick-start my writing as a visiting scholar at Northwestern University during the summer of 2011. Ever generous, smart, and nice, Sarah Brouillette invited me to Carleton University as the 2012 Faculty of Arts and Sciences Visiting Scholar, and I spent a few spring weeks in Ottawa giving talks about Curie and getting the chance to read and work on my book. While in the Canadian capital, I benefited greatly from meeting Michael Geist, Eli MacLaren, and Laura Murray, and had nice chats with Sara Bannerman and Sheryl Hamilton. Thank you also to Sarah de Rijcke at the Centre for Science and Technology Studies at Leiden University for providing me with an opportunity to present my work at a research seminar in May 2012.

Most of the reading and writing, however, took place in Paris, my favorite city, and at BnF Tolbiac, my favorite library. First I acknowledge the help of the staff at the BnF, especially Michèle Sacquin (now at the Institut de France) and Guillaume Fau. Nathalie Huchette, Anaïs Massiot, and Natalie Pigeard at the Musée Curie Historical Resources Center all helped with documents and images and offered prompt and kind answers to my questions. During the spring of 2013 I had the privilege of giving two talks at the Collège de France. I want to thank Professor John Scheid for sponsoring my visit and Guillaume Kasperski for all his assistance during that period. I remember with great warmth the wonderful stay at the Fondation Hugot and the kindness and hospitality of Laurinda Nunesa. Thanks also to Lucie Campos and Gabriel Galvez-Behar. All who have helped transform this text into a book at the University of Chicago Press have been nothing but a pleasure to work with, but I particularly want to single out the professionalism and kindness I always received from Karen Merikangas Darling and Sophie Wereley, as well as the brilliant job done by my copy editor, Susan Tarcov, and my indexer, Mary Newberry.

Annika Olsson was the first person I called when the idea for this book began to take shape and the one who—during that same phone conversation—discovered the 1911 duels on YouTube. One of the best and most perceptive readers I know, she is always—thankfully—on standby to spot inconsistencies and weakness in a text. So is my husband Per, with whom I have shared more than thirty years of everyday life made up of countless not so everyday moments. Let's continue that way a few more decades, preferably spending as much time as possible in Paris. And finally, I don't quite know how it happened, but the three daughters playacting Blossom, Bubbles, and Buttercup when I wrote my first book fifteen years ago have turned into three independent and beautiful young women. Minna, Rebecca, and Aurora, my own Powerpuff Girls, you still save my day. And you always will.

Notes

Every effort has been made to keep the notes to a minimum. The bibliographic essay describes both primary sources and secondary sources in detail. In the case of websites, newspapers, journal articles, and official documents such as those of the League of Nations and the CICI, I use the full reference in the note. This goes also for the most substantial primary source in the book, the Fonds Curie or the Curie archives at the Bibliothèque national de France. These are abbreviated as NAF (Nouvelles acquisitions françaises) *côtes* 18365–18517. References in the endnotes follow the format of NAF 18450, followed by f. (for "feuille") indicating the page number within a particular *côte*. Unless indicated otherwise, all translations from French are by me. In the case of existing English-language editions, for instance for *Pierre Curie* and *Madame Curie*, I have used the English version but also provided page numbers for the original French text, indicated by [F] for French and [E] for English. Documents from the League of Nations exist in French and English versions, and as much as possible I have tried to cite directly from the English versions. Abbreviations in references to League of Nations documents are as follows: A—Assembly; C—CICI; 42(a)—minutes, type of document; 1921— year; XII—roman numerals assigned for intellectual cooperation.

INTRODUCTION

1. "Marie Curie Tops Poll of Most Inspirational Women in Science," UNESCO and *New Scientist* press release, July 1, 2009. http://www.unesco .org.uk/uploads/PR-PollofWomeninScience-1July2009.pdf, last accessed January 10, 2014.

2. "De Gaulle plus grand Français de tous le temps," *Le Nouvel Observateur*, April 6, 2005.

3. "Les héros des Français," *L'Histoire*, no. 242, April 2000, 31–39.

4. Lemire, *Marie Curie*, 203.

5. Merton, "Matthew Effect," 620.

6. Lemire, *Marie Curie*, 11 (Piaf and Cosette), 166 (sad Mary Poppins).

7. Roqué, "Marie Curie," 276.

8. Shapin, *Scientific Life*, xiii.

9. Dorothy Pomerantz, "Michael Jackson Leads Our List of the Top-Earning Dead Celebrities," *Forbes*, October 23, 2013. http://www.forbes.com/sites/dorothypomerantz/2013/10/23/michael-jackson-leads-our-list-of-the-top-earning-dead-celebrities/, last accessed January 10, 2014.

CHAPTER ONE

1. Curie, *Madame Curie*, [F] 310, [E] 222.

2. Paul Gauthier to Marie Curie, June 9, 1920, NAF 18450, f. 72.

3. Henry Neumann to Marie Curie, June 19, 1924, NAF 18459, f. 164.

4. Curie, *Pierre Curie*, [E] 111.

5. Curie, *Pierre Curie*, [F] 78.

6. Wilhelm Röntgen, "On a New Kind of Rays," *Nature* 1369 (1896): 274.

7. Listed in Jacques Danne, *Le radium: Sa préparation et ses propriétés* (Paris: Librairie Polytechnique Ch. Béranger, 1904), 75.

8. Curie, *Pierre Curie*, [E] 94, [F] 64.

9. Eugène Pouillet, *Traité theorique et pratique des brevets d'invention et de la contrefaçon* (Paris: Marchel et Billard, 1899), 121.

10. Louis Barthou, Sénat, séance du 7 juillet 1932, 1036.

11. Marie Curie to Jules Jeanneney, July 8, 1932, NAF 18453, f. 47.

12. Jules Jeanneney to Marie Curie, July 11, 1932, NAF 18453, f. 50.

13. Curie, *Pierre Curie*, [E] 82, [F] 54.

14. N.-M. Le Senne, *Droits et devoirs de la femme devant la loi française* (Paris: Hennuyer, 1884), 8.

15. Geison, *Private Science of Louis Pasteur*, 9.

16. *Carnets de la découverte* 1 (15 septembre 1897–18 mars 1898), 2 (18 mars–16 juillet 1898), and 3 (11 novembre 1898–12 juillet 1900), NAF 18379, 18380, 183781.

17. Barbo, *Pierre Curie*, 320n63. See also Hurwic, *Pierre Curie*, 152.

18. NAF 18434, ff. 151–54.

19. Mme Skłodowska Curie, "Rayons émis pas les composés de l'uranium et du thorium," *Comptes rendus* 126 (1898): 1101–3.

20. *Œuvres de Pierre Curie*, 615–18.

21. M. P. Curie et Mme S. Curie, "Sur une substance nouvelle radioactive, contenue dans la pechblende," *Comptes rendus* 127 (1898): 175.

22. M. Skłodowska Curie, *Recherches sur les substances radioactives* (Paris: Gauthier-Villars, 1904), 7; *Pierre Curie*, [E] 96, [F] 66.

23. M. P. Curie, Mme P. Curie and M. G. Bémont, "Sur une nouvelle substance fortement radio-active, contenue dans la pechblende," *Comptes rendus* 127 (1898): 1215.

24. Reid, *Marie Curie*, 91.

25. C.-G. Schmidt, "Sur les radiations émises par le thorium et ses composés," *Comptes rendus* 126 (1898): 1264.

26. Curie, *Recherches sur les substances radioactives*, 7.

27. Eugene Demarçay, "Sur le spectre d'une substance radio-active," *Comptes rendus* 127 (1898): 1218.

28. Henri de Parville, "Le radium," *Journal des débats politiques et littéraires*, December 24, 1903, 1–2.

29. Badash, "Radium, Radioactivity," 146.

30. Louis Olivier to Marie Curie, December 31, 1898, NAF 18459, ff. 273–74.

31. Skłodowska Curie, "Les rayons de Bequerel et le polonium," *Revue generale des sciences pures et appliquées* 2 (January 30, 1899), 41–50.

32. Curie, *Madame Curie*, [F] 226, [E] 160.

33. Irène Joliot-Curie, "Etude sur les carnets de laboratoire de la découverte du polonium et du radium," in *Pierre Curie* (Paris: Editions Odile Jacob, 1996), 138.

34. Curie, *Pierre Curie*, [E] 99, [F] 68.

35. Curie, *Pierre Curie*, [E] 101, [F] 70.

36. Mme Curie, "Sur le poids atomique du radium," *Comptes rendus* 135 (July 21, 1902): 161.

37. Lucier, "Court and Controversy," 154.

38. Curie, *Madame Curie*, [F] 283–86, [E] 203–5.

39. Geison, *Private Science of Louis Pasteur*, 41–42.

40. Barbo, *Pierre Curie*, 319n53.

41. Le Senne, *Droits et devoirs*, 22.

42. Le Senne, *Droits et devoirs*, 129.

43. Pouillet, *Traité*, 121.

44. Le Senne, *Droits et devoirs*, 129–30.

45. Jeanne Chauvin, *Proposition de loi sur la capacité des femmes mariées de disposer du produit de leur travail ou de leur industrie personnels* (Paris: Publications de l'Avant-Courrière, 1893), 2.

46. "Contrat de mariage." M. J. Ader, Notaire à Paris. 226, Boulevard Saint-Germain. Fonds BnF *côte* JV15.

47. Fuchs, "Magistrates and Mothers," 17.

48. Marie Curie to Frank S. Gray, October 10, 1913, NAF 18436, f. 192.

49. André Beaunier, "Pierre Curie," *Le Figaro*, April 20, 1906.

50. Barbo, *Pierre Curie*, 265.

51. NAF 18517, f. 20. "Cher Pierre que je ne reverrai plus . . ." (Journal 1906–7) by Marie Curie in *Pierre Curie* (Paris: Editions Odile Jacob, 1996), 182–83.

52. Adrienne Weill, "Marie Curie," in *Dictionary of Scientific Biography*, 2:500.

53. Curie, *Madame Curie*, [F] 269 and 352, [E] 192 and 254.

54. Quinn, *Marie Curie*, 231.

55. Georges Gouy to Marie Curie, May 8, 1906, NAF 18450, ff. 244–45. See also Blanc, *Pierre Curie: Correspondances*, 675–76.

56. Crosland, *Science under Control*, 227.

CHAPTER TWO

1. Pierre Curie to Christopher Aurivillius, November 19, 1903, NAF 18436, ff. 273–74.

2. Gaston Rouvier, "Un ménage des savants français," *Le Temps*, December 10, 1910.

3. Hamilton, *Impersonations*, 190.

4. *Les Illustres Modernes, ou Table de la vie privée des principaux personnages des deux Sexes, qui depuis la renaissance des Lettres, ont acquis de la célébrité en Europe, tant en Politique ou dans les Armées, que dans les Arts, les Sciences & la vie contemplative* (Paris: Leroy, 1788), 3.

5. Curie, *Pierre Curie*, [E] 133, [F] 96.

6. Pierre Curie to Georges Gouy, April 14, 1906, NAF 18515, ff. 32–34.

7. Curie, *Madame Curie*, [F] 320, [E] 229–30.

8. "Lauréats du Prix Nobel," *Le Petit Parisien*, December 10, 1911, 3.

9. Curie, *Madame Curie*, [F] 309, [E] 221.

10. Albert Laborde, "Pierre Curie dans son laboratoire," in *Conférence faite au Palais de la Découverte à l'occasion du 50e anniversaire de la mort*

de Pierre Curie (Paris: Les Conférences du Palais de la Découverte, 1956), 13.

11. Pierre Curie to Georges Gouy, January 22, 1904, NAF 18515, f. 13.

12. Open letter from Marie Curie dated November 18. *Le Temps*, November 20, 1910.

13. Bellanger, *Histoire générale*, 296; Charle, *Le siècle de la presse*, 160– 61.

14. Delporte, "Presse," 94.

15. Ch. Dauzats, "Les femmes à l'Institut," *Le Figaro*, December 1, 1910, 3.

16. "La candidature de Mme Curie soulève une têmpete," *Le Petit Parisien*, December 1, 1910, 2.

17. Gaston Darboux, "Mme Curie et l'Académie des sciences," *Le Temps*, December 31, 1910, 3.

18. See Blanc, *Pierre Curie: Correspondances*, 351–423, for the politics surrounding the 1903 Nobel Prize nomination.

19. "La discussion fut longue. L' ordre du jour ambigu," *Le Petit Parisien*, January 5, 1911, 2.

20. Bouvier, "L'Académie élira-t-elle Mme Curie?" *Le Matin*, January 4, 1911, 1.

21. Ch. Dauzats, "La séance de demain," *Le Figaro*, January 3, 1911, 3.

22. Gaston Bonnier cited in "L'Institut de France décide de respecter les traditions," *Le Matin*, January 5, 1911, 1.

23. "Mme Curie sera-t-elle élue?" *Le Matin*, January 10, 1911, 1.

24. Curie, *Madame Curie*, [F] 382, [E] 277.

25. "L'Académie des sciences examine aujourd'hui la candidature de Mme Curie," *L'Excelsior*, January 9, 1911.

26. "Le siège de M. Gernez est déclaré vacant," *Le Petit Parisien*, January 10, 1911, 2.

27. "Un tournoi académique: Une femme entrera-t-elle à l'Institut?" *L'Excelsior*, January 23, 1911.

28. Pierre L'Ermite, "Branly, Curie . . . ?" *La Croix*, January 22, 1911.

29. D'Arsonval, "L'Académie des sciences a élu Branly," *Le Matin*, January 24, 1911, 1.

30. "A l'Académie des sciences," *Le Temps*, January 25, 1911, 1.

31. "M. Branly l'emporte sur Mme Curie," *Le Petit Parisien*, January 24, 1911, 1.

32. Ch. Dauzats, "Une nouvelle candidature de Mme Curie?" *Le Figaro*, January 26, 1911, 3.

33. Curie, "Autobiographical Notes," 202–3.

34. NAF 18383, ff. 120–21.

35. "Le photographe est sans pitié," *Femina*, February 1, 1911, 57.

36. Marbo, *A travers deux siècles*, 111.

37. "L'aventure de Mme Curie et de M. Langevin," *Le Petit Journal*, November 5, 1911, 1–2.

38. "Mme Curie et M. Langevin," *Le Temps*, November 5, 1911, 7.

39. "Mme Curie et M. Langevin," open letter from Marie Curie, *Le Temps*, November 8, 1911, 7.

40. Fernand Hauser to Marie Curie, reprinted in *Le Temps*, November 8, 1911, 7.

41. Loi du 29 juillet 1881, in Georges Barbier, *Code expliqué de la presse: Traité général de la police de la presse et des délits de publication* (Paris: Marchal et Billard, 1887), 2:12–76.

42. "Une protestation," *L'Aurore*, January 14, 1898.

43. Hurwic, *Pierre Curie*, 131; Curie, *Madame Curie*, [F] 202; Charle and Telkes, *Les professeurs*, 93.

44. Léon Daudet, "Dreyfus contre Branly," *L'Action Française*, January 23, 1911, 1.

45. Harris, *Man on Devil's Island*, 245.

46. Daudet, "La science et la vertu," *L'Action Française*, November 6, 1911, 1.

47. Daudet, "Le rôle de Jean Dupuy," *L'Action Française*, November 20, 1911, 1.

48. Léon Bailby, "Pour Monsieur X . . . ," *L'Intransigeant*, November 6, 1911, 1.

49. Maurice Pujo, "Pour une mère," *L'Action Française*, November 18, 1911, 1.

50. Reid, *Marie Curie*, 200.

51. Gustave Téry, "Pour une mère," *L'Œuvre*, November 23, 1911. The letter is also reprinted in Blanc, *Marie Curie et le Nobel*, 63–69.

52. Daudet, "Avertissement à Maîtres Millerand et Poincaré," *L'Action Française*, November 17, 1911, 1.

53. Laborde, "Pierre Curie dans son laboratoire," 21.

54. Téry, "Pour une mère," 6.

55. Pujo, "Pour une mère," *L'Action Française*, November 23, 1911, 1.

56. Berenson, *Trial de Madame Caillaux*, 241.

57. Pujo, "Pour une mère: Les documents sont publiés," *L'Action Française*, November 22, 1911, 1.

58. Daudet, *Bréviaire du journalisme* (Paris: Gallimard, 1936), 128.

59. Daudet, "Avertissement," 1.

60. Henri Chervet, "Les journalistes et M. Daudet," *Gil Blas*, November 18, 1911.

61. Chervet, "Pour M. Léon Daudet," *Gil Blas*, November 20, 1911.

62. Emile Louis Bruno Bruneau de Laborie, *Les lois du duel* (Paris: Manzi, Joyant, 1906), 43.

63. De Laborie, *Les lois du duel*, 123–26.

64. "Le duel Daudet-Chervet," *L'Action Française*, November 24, 1911, 1.

65. "1911 Epee Duel: Henri Chervet vs Leon Daudet," YouTube http://www.youtube.com/watch?v=4QlUwιkoItE, last accessed January 10, 2014.

66. "1911 Epee Duel: Pierre Mortier vs Gustave Tery," YouTube http://www.youtube.com/watch?v=rElNQuBvFeQ, last accessed January 10, 2014.

67. Berlanstein, "Historicizing," 75–76.

68. Frevert, *Men of Honour*, 182.

69. Marbo, *A travers deux siècles*, 117–18.

70. "Le duel hier," *Le Petit Journal*, November 26, 1911, 2.

71. Frevert, *Men of Honour*, 13.

72. Christopher Aurivillius to Marie Curie, November 25, 1911, NAF 18443, f. 290.

73. The correspondance between Svante Arrhenius and Marie Curie, dated December 1 and December 5, 1911, is cited in Blanc, *Marie Curie et le Nobel*, 113–16.

74. Nobel speech, December 11, 1911, available from http://www.nobel prize.org/nobel_prizes/chemistry/laureates/1911/marie-curie-lecture .html, last accessed January 10, 2014.

75. Eva Ramstedt to Marie Curie, December 17, 1911, NAF 18461, ff. 35–36.

76. Curie, *Madame Curie*, [F] 387, [E] 281.

CHAPTER THREE

The correspondance between Marie Curie (MC) and Missy Brown Meloney (MBM) is collected under one main *côte*, NAF 18457 (now digitized and available through Gallica). Unless stated otherwise, all subsequent notes refer to this *côte*.

1. Pierre Lafitte to Marie Curie, May 4, 1921, NAF 18454, ff. 36–38.

2. "A la gloire de la science française," *Je Sais Tout*, May 15, 1921, 569–72.

3. Hertha Ayron to Marie Curie, February 28, 1912, NAF 18443, f. 313.

4. Seventeen hits as applicant, fourteen as inventor, Espacenet http://www.epo.org/searching/free/espacenet.html, last accessed January 10, 2014.

5. Hertha Ayrton to Marie Curie, May 28, 1912, NAF 18443, f. 316.

6. Emmeline Pankhurst, *My Own Story* (London: Eveleigh Nash, 1914), 251.

7. Marie Curie to Henriette Perrin, May 10, 1921, NAF 18460, ff. 164–65

8. Meloney, introduction to *Pierre Curie*, [E] 16–17.

9. Henri Pierre Roché to Marie Curie, May 21, 1920, ff. 4–8.

10. MBM to MC, September 16, 1920, ff. 9–10.

11. MBM to MC, September 29, 1920, f. 11.

12. MC to MBM, November 7, 1920, f. 12.

13. MBM to MC, December 15, 1920, f. 13.

14. MBM to MC, January 4, 1921, f. 19.

15. MBM to MC, January 12, 1921, ff. 22–23.

16. MC to MBM, November 7, 1920, f. 12.

17. MBM to MC, January 24, 1921, ff. 24–25.

18. MBM to MC, December 28, 1920, f. 17.

19. MC to MBM, January 31, 1921, ff. 28–29.

20. MBM to MC, January 12, 1921, f. 22.

21. MBM to MC, February 17, 1921, ff. 39–40.

22. MC to MBM, March 9, 1921, ff. 51–52.

23. Paul Appell to Marie Curie, February 23, 1921, NAF 18443, ff. 176–77.

24. MBM to MC, March 23, 1921, ff. 55–56.

25. MBM to MC, April 5, 1921, f. 57.

26. "The Offer a Gift Well Earned (Topics of the Times)," *New York Times*, March 9, 1921.

27. "Public Notices," *New York Times*, April 21, 1921.

28. "To Supply Curie Radium," *New York Times*, May 4, 1921.

29. "Mme Curie Here Today," *New York Times*, May 11, 1921.

30. Curie, *Madame Curie*, [E] 326, [F] 446.

31. "Mme. Curie Plans to End All Cancers," *New York Times*, May 12, 1921.

32. "Radium Gift Awaits Mme. Curie Here," and "Mme. Curie Sails May 7," *New York Times*, February 7 and March 13, 1921.

33. Speech by President Warren Harding, May 20, 1921, NAF 18467, ff. 98–106.

34. Hyde, *The Gift*, 51.

35. Agreement between the executive committee of the Marie Curie Radium Fund and Madame Marie Curie, May 19, 1921, NAF 18467, f. 231 (handwritten), f. 259 (typed). Signed by notary public, f. 260.

36. "The Gift to Mme. Curie," *Science* 1379 (June 3, 1921): 513.

37. Curie, *Madame Curie*, [E] 331, [F] 452.

38. "Cancer Deaths Here Are Increasing," *New York Times*, March 7, 1921.

39. "A $100.000 Thimbleful," editorial, *Delineator*, June 1921, 1.

40. "A $100.000 Thimbleful," editorial, *Delineator*, June 1921, 1.

41. "Mme. Curie's Brain Fagged by 'Small Talk' of Americans," *New York Times*, May 28, 1921.

42. "Olympic Departs with 2,031 Aboard," *New York Times*, June 26, 1921.

43. Certificate of Insurance, Federal Insurance Company, June 11, 1921, no. 873020, NAF 18467, f. 242.

44. "Mme Curie, retour d'Amérique dit l'accueil enthousiaste qu'elle y reçut," *Le Petit Parisien*, July 3, 1921.

45. "Plan Life Income Now for Mme. Curie," *New York Times*, July 30, 1921.

46. MBM to MC, August 11, 1921, f. 69.

47. Minutes from meeting of the Marie Curie Radium Fund Committee, May 31, 1921, NAF 18467, f. 238.

48. MBM to MC, August 11, 1921, ff. 68–70.

49. MBM to MC, August 29, 1921, ff. 71–72.

50. MBM to MC, September 15, 1921, ff. 73–74

51. MC to MBM, September 29, 1921, f. 77.

52. Minutes from meeting with the subcommittee of the Marie Curie Radium Fund Committee, July 7, 1921, NAF 18467, f. 239.

53. Elsie Mead to Marie Curie, October 24, 1921, NAF 18456, f. 242.

54. MBM to Marie Curie, December 2, 1921, ff. 86–87.

55. Frederic R. Coudert to MBM, October 24, 1921, f. 88.

56. J. N. Babcock to MBM, November 17, 1921, f. 89.

57. MC to MBM, December 15, 1921, f. 91.

58. George E. Vincent, President, Rockefeller Foundation, to Marie Curie, November 18, 1921, NAF 18461, f. 169.

59. MC to MBM, March 17, 1922, f. 97.

60. MBM to MC, February 17, 1922, ff. 95–96.

61. MC to MBM, March 17, 1922, ff. 97–99.

62. MBM to MC, March 31, 1922, f. 102.

63. MBM to MC, April 25, 1922, f. 107.

64. MBM to MC, May 4, 1922, f. 111.

65. Letter to Marie Curie signed by twelve female chemists, Kent Chemical Laboratory, University of Chicago, January 27, 1921, NAF 18467, f. 156. Marie Curie to the University of Chicago, February 17, 1921, NAF 18467, f. 159.

66. MBM to MC, July 6, 1922, f. 117.

67. Ruth Beard Addis to MC, October 19, 1921, f. 81.

68. Publishing contract for *Pierre Curie*, NAF 18436, f. 365.

69. MBM to MC, February 17, 1922, ff. 95–96.

70. Marie Curie to Payot, December 19, 1921, NAF 18450, f. 78.

71. Contract with Macmillan copied in letter from MBM to MC, March 22, 1922, ff. 99–101.

72. MBM to MC, May 2, 1922, ff. 108–10.

73. MBM to MC, November 14, 1923, f. 155.

74. MC to MBM, October 6, 1922, ff. 118–19.

75. Curie, "Autobiographical Notes," 225.

76. NAF 18383, ff. 139, 174, 213–14.

77. MBM to MC, July 9, 1923, f. 139.

78. MC to MBM, May 30, 1923, f. 135.

79. *Journal Officiel*, January 18, 1924, NAF 18442, f. 11.

80. Curie, *Madame Curie*, [F] 351, [E] 252.

81. MBM to MC, December 15, 1923, NAF 18442, f. 180.

82. MBM to MC, December 27, 1923, ff. 166–66.

83. MC to MBM, January 4, 1924, ff. 168–69.

84. L. Razet to Macmillan, January 16, 1924, NAF 18456, f. 21.

85. Macmillan to L. Razet, February 9, 1924, NAF 18456, f. 22.

86. L. Razet to Macmillan, May 20, 1924, NAF 18456, f. 21.

87. Undated letter from Charlotte Kellogg to Marie Curie, reply from Marie Curie, January 4, 1924, NAF 18442, ff. 172–73.

88. MBM to MC, January 17, 1924, ff. 169–70.

89. MC to MBM, January 29, 1924, f. 171.

90. MC to MBM, April 22, 1924, f. 174.

91. Morey A. Park to Marie Curie, December 26, 1924, NAF 18458, f. 275.

92. Marie Curie (via L. Razet) to Morey Flux and Chemical Company, January 21, 1925, NAF 18458, f. 276.

93. Marie Curie to Richard B. Moore, January 21, 1925, NAF 18458, f. 274.

94. Richard B. Moore to Morey A. Park, February 18, 1925, NAF 18458, f. 279.

95. Morey A. Park to Richard B. Moore, February 26, 1925, NAF 18458, f. 280.

96. Richard B. Moore to Marie Curie, March 2, 1925, NAF 18458, f. 277.

97. MBM to MC, November 14, 1924, ff. 181–82.

98. MBM to MC, May 18, 1925, f. 200.

99. Marie Curie to William D. Coolidge, Director, Research Laboratory, General Electrical Company, January 8, 1934, NAF 18447, f. 157.

100. MBM to MC, September 8, 1927, f. 236.

101. MBM to MC, April 25, 1929, f. 271.

102. MC to MBM, January 27, 1930, f. 326.

103. Union minière to Owen D. Young, December 27, 1929, NAF 18465, ff. 133–34.

104. Roqué, "Marie Curie and the Radium Industry," 268.

105. Treasurer's Report, Madame Curie Radium Fund, September 4–October 31, 1929, NAF 18467, f. 281.

106. Stanford University Acting President Robert E. Swain to University of Paris Rector S. Charléty, January 25, 1930, MF 18456, f. 204.

107. MBM to MC, June 2, 1931, f. 398.

108. MC to MBM, July 18, 1932, f. 434.

109. MBM to MC, August 3, 1932, ff. 435–37.

110. MC to MBM, August 25(29?), 1932, f. 439.

111. MBM to MC, July 10, 1928, f. 256.

112. MBM to MC, September 25, 1929, f. 297–98.

113. MBM to MC, September 19, 1932, f. 446.

114. MBM to MC, October 28, 1932, f. 452.

115. MBM to MC, March 10, 1933, f. 464.

116. Marie Curie Radium Fund, undated resolution enclosed in letter from MBM to MC, March 10, 1933, f. 465. Copy of the resolution with signatures in NAF 18467, f. 262.

117. *Testament du radium*, March 25, 1934. Letter from Marie Curie. Archives du Musée Curie, AIR LC.MC.

118. MBM to MC, March 31, 1922, ff. 102–3.

119. MBM to MC, March 11, 1930, f. 340.

120. MBM to MC, September 17, 1930, f. 365.

121. MBM to MC, October 1, 1931 (f. 409); March 15, 1932 (f. 427); September 13, 1932 (f. 443); March 27, 1933 (f. 466); September 16, 1933 (f. 494).

122. MC to MBM, October 10, 1933, f. 499.

123. MBM to MC, June 6, 1934, f. 508.

CHAPTER FOUR

1. A. Laquerrière, *La Presse Medicale*, August 31, 1921, 1267.

2. Hertha Ayrton to Marie Curie, December 12, 1921, NAF 18443, f. 321.

3. Methuen & Co. to Félix Alcan, January 12, 1922, NAF 18443, f. 78.

4. Félix Alcan to Marie Curie, October 24, 1921, NAF 18443, f. 73.

5. Félix Alcan to Marie Curie, March 18, 1924, NAF 18443, f. 80.

6. Curie, *Madame Curie*, [E] 340, [F] 464.

7. Eric Drummond to Marie Curie, May 17, 1922, NAF 18463, ff. 3–5.

8. Albert Einstein to Marie Curie, September 6, 1929, NAF 18449, f. 63.

9. Marie Curie to Eric Drummond, June 1, 1922, NAF 18463, f. 9.

10. Eric Drummond to Marie Curie, June 7, 1922, NAF 18463, f. 11.

11. Marie Curie to Albert Einstein, January 6, 1924. Albert Einstein, *Correspondances françaises* (Paris: Seuil, 1989), 81.

12. A.61.C.559.1922.XII, 4.

13. A.61.C.559.1922.XII, 7.

14. A.61.C.559.1922.XII, 5.

15. Paul Otlet, *Sur l'établissement en Belgique du siège de la Société des Nations* (Brussels, 1919).

16. Cited in *Cent ans de l'Office international de bibliographie*, 39.

17. Cited in *Cent ans de l'Office international de bibliographie*, 39.

18. *Le Répertoire bibliographique universel et la coopération internationale dans les travaux bibliographiques*, Congrès Bibliographique International, Paris 1900 (Brussels, 1900), 5.

19. Henri La Fontaine and Paul Otlet, "Sur la création d'un Répertoire Bibliographique Universel," in *Conférence bibliographique internationale, Bruxelles 1895, Documents* (Brussels: Larcier, 1895), 6–7.

20. La Fontaine and Otlet, "Sur la création," 8.

21. Letter of invitation dated July 30, 1895, printed in *Conférence bibliographique internationale, Bruxelles 1895, Documents*.

22. La Fontaine and Otlet, "Sur la création," 17.

23. La Fontaine and Otlet, "Sur la création," 19.

24. *Le Répertoire bibliographique universel*, 4.

25. Paul Otlet, *Le mouvement scientifique en Belgique, 1830–1905* (Brussels: L'Office international de bibliographie, 1910), 9.

26. *Rapports présentés au Congrès international de physique réuni à Paris en 1900 sous les auspices de la Société française de physique* (Paris: Gauthier-Villars, 1900–1901), 4:53.

27. A.42(b).1921, 3.

28. Union des associations internationales, *Organisation internationale du travail intellectuel*, UAI publication no. 97 (Brussels, June 1921), 7.

29. *La Propriété Industrielle*, June 30, 1922, 87–88.

30. Henri Weindel, *Communications sur les origines, les buts et les moyens d'action de la Confédération des travailleurs intellectuels*, presented to the Académie des sciences morales et politiques, October 16, 1920 (Paris: Hémery, n.d.), 3.

31. Marbo, *A travers deux siècles*, 113.

32. Weindel, *Communications*, 13.

33. "Lauréats du Prix Nobel," *Le Petit Parisien*, December 10, 1911, 3.

34. NAF 18452, f. 203–4.

35. Letter from Marie Curie to the Union des Associations d'Ancièns Elèves des Ecoles de Chimie, NAF 18452, f. 205.

36. MBM to MC, June 24, 1925, f. 202.

37. MC to MBM, July 8, 1925, f. 203.

38. Florence L. Pfalzgraf to Marie Curie, May 25, 1928, NAF 18444, ff. 198–99.

39. Roger Dalimier and Louis Gallié, *La propriété scientifique: Le projet de la C.T.I. Création d'un droit d'auteur pour le savant et l'inventeur* (Paris: A. Rousseau, 1923), 71.

40. Boudia, *Marie Curie et son laboratoire*, 89; Hurwic, *Pierre Curie*, 132.

41. A.38.1923.XII, 1.

42. Borel, preface to Dalimier and Gallié, *La propriété scientifique*, 16.

43. A.38.1923.XII, 6.

44. A.38.1923.XII, 14.

45. Davies, *Property*, 16.

46. A.38.1923.XII. Annexe, 26.

47. Cesare Vivante, "La propriété scientifique devant la Société des nations," *Annales de droit commercial français, étranger et international* (1924), 2.

48. J. David Thompson to Marie Curie, May 3, 1928, NAF 18463, f. 73.

49. Vernon Kellogg to J. David Thompson, May 2, 1928, NAF 18463, f. 74.

50. A.29.1924.XII, 5.

51. C.3.M.3.1924.XII, 21.

52. "L'inauguration de l'Institut de cooperation intellectuelle," *Le Temps*, January 17, 1926.

53. Marie Curie to Vernon Kellogg, March 30, 1923, NAF 18463, f. 28.

54. Marie Curie to Shrohl & Science Abstracts, NAF 18463, ff. 16–17.

55. Levie, *L'homme qui voulait classer le monde*, 200.

56. A.31.1923.XII, 5.

57. NAF 18441, f. 144.

58. NAF 18441, f. 6.

59. Bensaude-Vincent, *Langevin*, 90–94.

60. Transcripts from meetings and various versions of the draft law in NAF 18441, ff. 233–337.

61. MBM to MC, February 2, 1932, ff. 425–26.

62. Jules Destrée, "La crise de la Commission internationale de cooperation intellectuelle," *Le Soir*, August 16, 1929.

63. Madame Razet to Mr. Hooper, *Encyclopedia Britannica*, May 6, 1924, NAF 18449, f. 110.

64. Madame Razet to *Encyclopedia Britannica*, June 18, 1924, NAF 18449, f. 117.

65. MBM to MC, May 17, 1926, f. 216.

66. MC to MBM, May 29, 1926, f. 220.

67. MC to MBM, April 30, 1926, f. 214.

68. MC to MBM, January 20, 1930, f. 322.

69. MC to MBM, January 25, 1930, f. 324.

70. P. J. Philip to Marie Curie, *New York Times*, January 7, 1930, NAF 18459, f. 195.

71. Marie Curie to P. J. Philip, January 15, 1930, NAF 18459, f. 197.

72. MBM to MC, February 7, 1930, f. 327.

73. Cecil Hunt to Marie Curie, November 4, 1930, NAF 18451, f. 243.

74. William Albert Lorenz to Marie Curie, March 19, 1924, NAF 18455, f. 231.

75. Charles Eugene Claghorn to Marie Curie, February 27 and April 15, 1933. Madame Razet to Claghorn, April 27, 1933. NAF 18446, ff. 191–93.

76. Francis Laur, "Une interview de Mme Curie," *Les Inventions Illustrées*, March 5, 1911.

77. Letter from Madame Razet, February 12, 1924, NAF 18445, f. 252, in reply to a request from Milton Bronner, Newspaper Enterprise Association of America, dated February 6, 1924, NAF 18445, f. 251.

78. A.38.1923.XII, 20.

79. George K. Burgess to Marie Curie, May 23, 1923, NAF 18445, ff. 293–94.

80. George K. Burgess to Marie Curie, June 2, 1923, NAF 18445, f. 295.

81. Marie Curie to George K. Burgess, June 19, 1923, NAF 18445, f. 296.

82. Cassier and Sinding, "'Patenting in the Public Interest,'" 155.

83. T. Swann Harding, "Exploitation of the Creators," *Philosophy of Science* 3 (1941): 386.

84. Lefebvre and Raynal, "De l'Institut Pasteur à Radio Luxembourg," 462.

85. Letter from J. L. de Ricqlès, February 2, 1934, NAF 18447, f. 293.

86. Paul Allard, "La propriété scientifique," *L'Excelsior*, December 29, 1929, 4.

87. NAF 18443, f. 86.

88. Marie Curie, *rapporteur*, "Rapport sur la question de la propriété scientifique," séance du 23 juin 1931, *Bulletin de l'Académie de médecine* 105 (1931):982–86.

89. NAF 18441, f. 91.

90. Lucien Klotz, "Les droits de la science: Pour les œuvres utiles à la santé publique il faut des millions. La loi sur la propriété scientifique les fournira," séance du 21 avril 1931, *Bulletin de l'Academie de médecine* 105 (1931): 678–85.

91. Ernest Forneau, "Discussion sur le rapport de Mme Curie sur la question de la propriété scientifique," séance du 7 juillet 1931, *Bulletin de l'Académie de médecine* 106 (1931): 3–9.

92. A. 21.1930.XII, 4.

93. *L'avenir de la culture* (Paris: Institut international de coopération intellectuelle, 1933), 209–10.

94. *L'avenir de la culture*, 11–25.

95. *L'avenir de la culture*, 214.

96. Marie Curie to Albert Einstein, August 30, 1929, NAF 18449, f. 62.

97. NAF 18441, f. 141.

98. Union des associations internationales, *Centre Internationale*, publication no. 98 (Brussels: August 1921), 5.

99. Eve Curie to Dr. Tobé, June 22, 1934, NAF 18435, ff. 168–69.

100. Dr. Tobé to Eve Curie, June 23, 1934, NAF 18435, f. 172.

101. Death certificate by Dr. Tobé, July 4, 1934, NAF 18435, f. 184.

102. "Sancellemoz, dernières paroles de Madame Curie," NAF 18435, f. 187.

103. Eric Pfanner, "Google to Announce Venture with Belgian Museum," *New York Times*, March 12, 2012. http://www.nytimes.com /2012/03/13/technology/google-to-announce-venture-with-belgian -museum.html?_r=0, last accessed January 10, 2014.

EPILOGUE

1. Running between September 4 and October 23, 1937.

2. "Greta Garbo to Portray Role of 'Madame Curie,'" *Herald-Journal*, October 2, 1938.

3. MC to MBM, May 20, 1922, f. 112.

4. George Sarton, review of *Madame Curie, Pierre Curie*, and *Marie Skłodowska-Curie, 1867–1934*, by Claudius Regaud, *Isis* 2 (1938): 480–84.

5. Pinault, "Marie Curie," 11.

6. Marbo, *A travers deux siècles*, 121–22.

7. George R. Manue to Marie Curie, September 1, 1932, NAF 18456, f. 96.

8. MC to MBM, November 28, 1929, f. 308.

9. MBM to MC, April 1, 1931, f. 388.

10. ALI to Marie Curie, August 5, 1930, NAF 18462, ff. 111–12.

11. Marie Curie (via Madame Razet) to ALI, August 27, 1930, NAF 18462, f. 113.

12. Stanley Unwin to Géo Robert, December 30, 1930, NAF 18461, f. 145.

13. Marie Curie to Géo Robert, January 12, 1931, NAF 18461, f. 146.

14. Methuen & Co. to Marie Curie, 5 November 1924, and reply from Marie Curie, November 10, 1924, NAF 18455, ff. 251–52.

15. MBM to MC, January 22, 1930, f. 323.

16. MBM to MC, July 1, 1929, f. 278.

17. MBM to MC, December 18, 1924, f. 185.

18. Marie Curie to Mrs. Edsel Ford, February 18, 1930, NAF 18449, f. 276.

19. Handwritten note by Irène Joliot-Curie, August 15, 1941 [copie dossiers CEA 1 à 20]. Consulted at the Archives du Musée Curie, July 16, 2013.

20. Marie Curie, "The Discovery of Radium," address at Vassar College, May 14, 1921, available from http://gos.sbc.edu/c/curie1921.html, last accessed January 10, 2014.

21. "Marie Skłodowska Curie," *Scientific American*, November 25, 1911, 471.

22. Curie, *Madame Curie*, [F] 245, [E] 173.

23. Address of Dr. Richard B. Moore, Chief Chemist, United States Bureau of Mines, May 17, 1921, NAF 18467, f. 36.

24. P. Curie, M. Curie, and G. Bémont, "Radium, a New Body, Strongly Radio-active, Contained in Pitchblende," *Scientific American*, January 28, 1899, 60.

25. François Mitterand, "Discours du transfert des cendres de Pierre et Marie Curie au Panthéon," April 20, 1995, available from http://fr.wikisource.org/wiki/Discours_du_transfert_des_cendres_de_Pierre_et_Marie_Curie_au_Panthéon, last accessed January 10, 2014.

26. "Ceremonie Panthéon," available from Institut national audio-visuel (INA), http://www.ina.fr/video/CAB95027036/ceremonie-pantheon-video.html, last accessed January 10, 2014.

27. *L'avenir de la culture* (Paris: Institut international de coopération intellectuelle, 1933), 250.

28. Curie, *Pierre Curie*, [E] 100, [F] 69.

29. Shapin, *Scientific Life*, 19.

30. See the ESPCI website http://www.espci.fr/fr/espci-paristech/, last accessed January 10, 2014.

31. NAF 18441, f. 313.

32. Union des associations internationales, *Organisation internationale du travail intellectuel*, UAI publication no. 97 (Brussels, June 1921), 6.

33. Press release, Curie Institute, September 12, 2001, http://curie.fr/sites/default/files/myriadopposition6sept01_gb.pdf, last accessed January 10, 2014.

34. Hanna Stein to Marie Curie, July 23, 1925, NAF 18463, f. 337.

35. Letter from the *Gazette Apicole Montfavet-Avignon*, October 30,

1933, reply from Madame Razet, November 2, 1933, NAF 18450, ff. 94–96.

36. Parfumerie Bourjois, New York, to Marie Curie, October 27, 1931, NAF 18445, f. 110. Cable from Curie to Parfumerie Bourjois, November 7, 1931.

37. R. Cortesi to Marie Curie, December 14, 1933; Madame Razet to R. Cortesi, January 9, 1934, NAF 18447, ff. 180–81.

38. Correspondance between H. Becker and Marie Curie, NAF 18444, ff. 87–91.

39. NAF 18454, f. 54

40. Frow, "Elvis' Fame," 139.

41. For a good overview of the relative merits of the case and the various twists and turns, see Roger V. Skalbeck, "How Dewey Classify OCLC's Lawsuit," LLRX, September 29, 2003, http://www.llrx.com /features/deweyoclc.htm, last accessed January 10, 2014.

42. Press release, OCLC, November 24, 2003, available from http:// worldcat.org/arcviewer/2/OCC/2010/05/07/H1273247357646/viewer /file581.htm, last accessed January 10, 2014.

43. Litman, "Breakfast with Batman," 1722.

44. "MCA on Facebook," http://ec.europa.eu/research/mariecurie actions/news-events/news/2013/mca_on_facebook_en.htm, last accessed January 10, 2014.

45. MBM to MC, August 22, 1929, f. 288.

Bibliographic Essay

Following Ginette Gablot's "A Parisian Walk along the Landmarks of the Discovery of Radioactivity," *Physics in Perspective* 2 (2000): 100–107, will inevitably take you to the modest Curie Museum. Although you might be content simply perusing the permanent exhibition, the Musée Curie Historical Resources Center, with its archives primarily related to the Radium Institute and the Fondation Curie (http://curie.fr/en/fondation/curie-museum), is only a block or so away. But it was elsewhere in the city that I consulted the most important primary source for this book. The best way to get a feel for the scope of the Pierre et Marie Curie Papiers (NAF 18365–18517) at the Bibliothèque national de France (BnF) (http://www.bnf.fr) is by consulting the collection index available as a pdf document via the BnF Archives and Manuscripts department (http://archivesetmanuscrits.bnf.fr/ead.html?id=FRBNFEAD000007376&c=FRBNFEAD000007376_e0000018&qid=sdx_q13). However, because the BnF continuously migrated many of the 152 *côtes* into their incomparable digital library Gallica during the period I spent working on this project, significant parts of the collection are today available from the Gallica website (http://gallica.bnf.fr), not only making the horrors of microfilm consultation a thing of the past, but turning my iPad mini into a portable Curie archive. Where Gallica is at its most impressive, however, is in providing access to French newspapers, setting an example in making available one of the most important resources for any historian of the late modern period. Much of Pierre and Marie Curie's own writing can also be accessed via Gallica's digitized scholarly journals, of which the *Comptes rendus* is the best known. Pierre Curie's collected works, published (with an introduction by his widow) in *Œuvres de Pierre Curie* (Paris: Gauthier-Villars, 1908 [reprinted 1984]), can be found there as well, as can the Payot edition of Marie Curie's *Pierre Curie*. The troubled American edition from Macmillan, including Curie's "Autobiographical Notes," is more difficult to find in libraries and has been consulted here in an electronic version, http://etext.virginia.edu/toc/modeng/public /CurPier.html. Interestingly enough, there has been no French publication of Marie Curie's complete works. Prefaced by Irène Joliot-Curie, *Œuvres de Marie Skłodowska Curie* (Warsaw: Academie Polonaise des Sciences, 1954) was published in Warsaw on the twentieth anniversary of her mother's death. Karin Blanc's *Pierre Curie: Correspondances* (Paris: Hayot, 2009) is well structured and informative, but aside from *Lettres: Marie Curie et ses filles* (Paris: Pygmalion, 2010), a fairly recent selection of letters exchanged between Curie and her daughters, there is no comparable volume on what remains of Marie Curie's personal and professional correspondence.

INTRODUCTION

Even though this is not a biography in the traditional sense of the word, it might be useful to start with a few examples in that genre. Because it has been formative in shaping the Curie myth, Eve Curie's *Madame Curie* (Paris: Gallimard, 2010 [1938]), remains requisite reading for anyone interested in Curie's trajectory as a person and persona. I have relied on the French Folio edition and the American DaCapo edition from 1982, which is an unabridged version of the original Doubleday book from 1937. Susan Quinn, *Marie Curie: A Life* (Cambridge, MA: DaCapo Press, 1995), is the most authoritative biography to date, with Robert Reid, *Marie Curie* (London: Collins, 1974), still a very readable option. Many subsequent biographies are heavily indebted to these two books and bring little new to the table. Instead, one can start with Quinn and Reid and move on to Lauren Redniss's beautiful and innovative *Radioactive: Marie and Pierre Curie, a Tale of Love and Fallout* (New York: It Books, 2010) or browse Françoise Balibar's richly illustrated *Marie Curie: Femme savante ou Sainte Vierge de la science?* (Paris: Gallimard, 2006). Biographies of Pierre Curie include Anna Hurwic's *Pierre Curie* (Paris:

Flammarion, 1998) and Loïc Barbo's *Pierre Curie 1859–1906: Le rêve scientifique* (Paris: Belin, 1999). The extended Curie clan is at the center of attention of Eugenie Cotton, *Les Curie* (Paris: Éditions Seghers, 1963); Brian Denis, *The Curies: A Biography of the Most Controversial Family in Science* (New York: Wiley, 2005); and Pierre Radvanyi, *Les Curie: Pionniers de l'atome* (Paris: Belin, 2005). Camille Marbo, *A travers deux siècles: Souvenirs et rencontres, 1883–1967* (Paris: Grasset, 1967), is one of the few firsthand accounts of Curie's and her family's private life. Shelley Emling's *Marie Curie and Her Daughters: The Private Lives of Science's First Family* (New York: Palgrave Macmillan, 2012) is a popular account detailing the relationship between Curie and her daughters.

An alternative way of exploring Curie's biographical fate is by considering what strikes me as quite different French and English traditions when it comes to writing biography. From Françoise Giroud's bestseller *Une femme honorable* (Paris: Fayard, 1981) to Laurent Lemire's comparisons with Cosette, Edit Piaf, and Mary Poppins in *Marie Curie* (Paris: Perrin, 2001) over to *Marie Curie* by Henri Gidel (Paris: Flammarion, 2008) and Janine Troterau's more recent book with the same name (Paris: Gallimard, 2011), French biographies tend toward the fictional, drawing on a dramatic and emotional language that is far less common in English equivalents (although by no means completely absent). Laurent Lemire's comment that the depreciation of Curie might be the result of an Anglo-American failure to understand the French way of doing science is a reminder of just how much national investment goes into the construction of scientists (and science). On that score, see for instance Marjorie Malley, "The Discovery of Atomic Transmutation: Scientific Styles and Philosophies in France and Britain," *Isis* 70, no. 2 (1979): 213–23, and Mary Jo Nye, "National Styles? French and English Chemistry in the Nineteenth and Early Twentieth Centuries," *Osiris*, 2nd ser., 8 (1993): 30–49. Contributions weighing in on the importance of biography as part of the history of science include Mary Jo Nye, "Scientific Biography: History of Science by Another Means?" *Isis* 97, no. 2 (2006): 322–29; Mott T. Greene, "Writing Scientific Biography," *Journal of the History of Biology* 40, no. 4 (2007): 727–59; Mary Terrall, "Biography as Cultural History of Science," *Isis* 97, no. 2 (2006): 306–313, and anthologies such as Michael Shortland and Richard Yeo, eds., *Telling Lives in Science: Essays on Scientific Biography* (Cambridge: Cambridge University Press, 1996), and

Tomas Söderqvist, ed., *The History and Poetics of Scientific Biography* (Aldershot, UK: Ashgate, 2007).

In recent years, however, a new type of "biography" has emerged, and it is one to which this present book owes a greater debt. What we could call metabiographies may concern themselves with individuals, but their goal is to say something about the broader influences, tendencies, and trajectories illustrated by a scientist like Curie. Maria Rentetzi's creative biography of radium, *Trafficking Materials and Gendered Experimental Practices: Radium Research in Early 20th Century Vienna* (New York: Columbia University Press, 2008) comes to mind, as does Nicolaas Rupke, *Alexander von Humboldt: A Metabiography* (Chicago: University of Chicago Press, 2008); Mario Biagioli, *Galileo's Instruments of Credit: Telescopes, Images, Secrecy* (Chicago: University of Chicago Press, 2006); Steven Shapin, *The Scientific Life: A Moral History of a Late Modern Vocation* (Chicago: University of Chicago Press, 2008); and, of course, Bruno Latour's *Pasteur: Guerre et paix des microbes* (Paris: La Découverte, 2001). Although I would not place Edward Berenson's wonderful *The Trial of Madame Caillaux* (Berkeley: University of California Press, 1992) in the above category, I find its microhistorical perspective and narrative structure very appealing. For a good historiographical overview of a movement he has been closely associated with, see Carlo Ginzburg, "Microhistory: Two or Three Things That I Know about It," *Critical Inquiry* 20, no. 1 (1993): 10–35, and on microhistory as part of the history of science, Paula Findlen, "The Two Cultures of Scholarship?" *Isis* 96, no. 2 (2005): 230–37. Because I consider myself first a scholar of intellectual property and only second as belonging to a particular discipline, my greatest debt lies within a highly amorphous and multidisciplinary field of study that has exploded over the last twenty years. I can acknowledge only a few inspirations: Martha Woodmansee, *The Author, Art, and the Market* (New York: Columbia University Press, 1994); James Boyle, *Shamans, Software, and Spleens* (Cambridge, MA: Harvard University Press, 1996); Jane Gaines, *Contested Culture* (Chapel Hill: University of North Carolina Press, 1999); Eli MacLaren, *Dominion and Agency* (Toronto: University of Toronto Press, 2011); Adrian Johns, *Piracy* (Chicago: University of Chicago Press, 2009); Carol Rose, *Property and Persuasion* (Boulder, CO: Westview Press, 1994). Anthologies include Martha Woodmansee and Peter Jaszi, eds., *The Construction of Authorship* (Durham, NC: Duke University Press, 1999), and Mario Biagioli et al., *Making and Unmaking Intellectual Property* (Chicago: University of Chicago Press, 2011).

CHAPTER ONE

Robert Merton discusses disinterestedness in *Social Theory and Social Structure* (New York: Free Press, 1968). Merton has also written on intellectual property in "The Matthew Effect in Science, II: Cumulative Advantage and the Symbolism of Intellectual Property," *Isis* 79, no. 4 (1988): 606–23. My overall framing of the work going into the separation of science from nonscience relies substantially on Thomas Gieryn, *Cultural Boundaries of Science: Credibility on the Line* (Chicago: University of Chicago Press, 1999).

On the general conditions for women in French universities at the time Curie began her career, see Jean-François Condette, "'Les Cervelines' ou les femmes indésirables: L'étudiante dans la France des années 1880–1914," *Carrefours de l'éducation* 15 (2003): 38–61. For a broader international backdrop for the general institutional conditions for science—pure as well as applied—during the period I am concerned with in this book, see Mary Jo Nye's *Before Big Science: The Pursuit of Modern Chemistry and Physics, 1800–1940* (Cambridge, MA: Harvard University Press, 1999); Daniel Kevles, *The Physicists: The History of a Scientific Community in Modern America* (Cambridge, MA: Harvard University Press, 1995); Robert Fox and Anna Guagnini, "Laboratories, Workshops and Sites: Concepts and Practices of Research in Industrial Europe, 1800–1914," *Historical Studies in the Physical and Biological Sciences* 29, no. 1 (1998): 55–139; and Paul Forman, John L. Heilbron and Spencer Weart, *Physics circa 1900: Personnel, Funding and Productivity of the Academic Establishments* (Princeton, NJ: Princeton University Press, 1975). Robert Fox's *The Savant and the State: Science and Cultural Politics in Nineteenth-Century France* (Baltimore, MD: Johns Hopkins University Press, 2012) synthesizes his decades of research on French science within the concept of the long nineteenth century.

Soraya Boudia's work remains indispensable when considering Marie Curie's own laboratory strategies, a less explored topic than one might perhaps think. See her *Marie Curie et son laboratoire: Sciences et industrie de la radioactivité en France* (Paris: Editions des archives contemporaines, 2001), and for more on the negotiations surrounding the radium standard described in chapter 2, "The Curie Laboratory: Radioactivity and Metrology," *History and Technology* 13 (1997): 249–65. This special issue of the journal dedicated to the extended Curie family also includes an interesting contribution by Xavier Roqué, "Marie Curie and the Radium Industry: A Preliminary Sketch" (267–91). Together with Boudia's texts, Roqué's "Displacing Radioactivity," in *Instrumentation: Between Science, State and Industry*, ed. Bernward Joerges and Terry Shinn (Amsterdam: Kluwer, 2001), 51–68, and J. L. Davis, "The Research School of Marie Curie in the Paris Faculty, 1907–14," *Annals of Science* 52 (1995): 321–55, all provide a more nuanced and complex image of Curie as an institution builder.

Obviously, the Code Civil enforced at the time in question has been revised many times over. I have relied on the 1804 original available via Gallica, but the most recent edition is *Code Civil* (Paris: Dalloz, 2013). Helpful general introductions include Jean-Louis Halpérin, *Histoire du Droit privé français depuis 1804* (Paris: PUF, 2001); Bernhard Schnapper, "Autorité domestique et partis politiques, de Napoléon à De Gaulle," in *Voies nouvelles en histoire du droit. La justice, la famille, la répression pénale. xvième–xxème siècles* (Paris: PUF, 1991): 555–96, and Romuald Szramkiewicz and Jacques Bouineau, *Histoire des institutions, 1750–1914: Droit et société en France de la fin de l'ancien régime à la première guerre mondiale* (Paris: Librairie de la Cour de cassation, 1989).

My reading of persona and property is heavily indebted to scholars in French history and feminist thought, as well as to a strand of legal scholarship considering the relation between the law and property. Excellent starting points for understanding the role of French women during the period in question are Christopher E. Forth and Elinor Accampo, eds., *Confronting Modernity in Fin-de-Siècle France: Bodies, Minds and Gender* (London: Palgrave Macmillan, 2010), and Joan Wallach Scott's *Only Paradoxes to Offer: French Feminists and the Rights of Man* (Cambridge, MA: Harvard University Press, 1996). On marriage and divorce, see Jean Pedersen, *Legislating the French Family* (New Brunswick, NJ: Rutgers University Press, 2003), and on paternity suits in court, Rachel Fuchs, "Magistrates and Mothers, Paternity and Property in Nineteenth-Century French Courts," *Crime, History and Societies* 13, no. 2 (2009): 13–26. Legal scholars whose work has been particularly relevant to understanding the relation between person and intellectual property include Sheryl N. Hamilton, *Impersonations: Troubling the Person in Law and Culture* (Toronto: University of Toronto Press, 2009), and Margaret Davies, *Property: Meanings, Histories, Theories* (Oxford:

Routledge, 2007). I especially enjoyed Ngaire Naffine's writings because they led me to the term "sexing." See "Our Legal Lives as Men, Women and Persons," *Legal Studies* 24 (2004): 621–42, and "Who Are Law's Persons? From Cheshire Cats to Responsible Subjects," *Modern Law Review* 66, no. 3 (2003): 346–67. Anne Lefebvre-Teillard, *Introduction historique au droit des personnes et de la famille* (Paris: PUF, 1996), provides a thorough overview of the phases through which name, family, and inheritance have traveled in continental legal thought.

Although scholarship on the relation between intellectual property and authorship has proliferated during the past twenty years, the interest in scientific authorship is perhaps more recent. Early studies on scientific collaboration and authorship include D. deB. Beaver and R. Rosen, "Studies in Scientific Collaboration, Part I: The Professional Origins of Scientific Co-authorship," *Scientometrics* 1 (1978): 65–84, with Mario Biagioli and Peter Galison, eds., *Scientific Authorship: Credit and Intellectual Property in Science* (London: Routledge, 2003), now established as a standard volume in the field. On the notebook as scientific genre and the private/public aspect, see Fredric L. Holmes, Jürgen Renn, and Hans-Jörg Rheinberger, *Reworking the Bench: Research Notebooks in the History of Science* (Secaucus, NJ: Kluwer Academic Publishers, 2003), and Gerald L Geison, *The Private Science of Louis Pasteur* (Princeton, NJ: Princeton University Press, 1995). Helena M. Pycior, Nancy G. Slack, and Pnina G. Abir-Am, eds., *Creative Couples in the Sciences* (New Brunswick, NJ: Rutgers University Press, 1996), and Annette Lykknes, Donald L. Opitz, and Brigitte Van Tiggelen, eds. *For Better or for Worse? Collaborative Couples in the Sciences* (Basel: Springer, 2012), both are concerned with collaboration between scientific couples. Helena M. Pycior, "Reaping the Benefits of Collaboration while Avoiding Its Pitfalls: Marie Curie's Rise to Scientific Prominence," *Social Studies of Science* 23, no. 2 (1993): 301–23, discusses the Curies' authorship strategies, and Margaret W. Rossiter, "The Matilda Effect in Science," *Social Studies of Science* 23, no. 2 (1993): 325–41, provides a highly readable analysis of the gendered mechanisms at play in collaboration.

Studies providing valuable background on the historical evolution of patent law include Pamela O. Long, *Openness, Secrecy, Authorship: Technical Arts and the Culture of Knowledge from Antiquity to the Renaissance* (Baltimore, MD: John Hopkins University Press, 2001); Christine MacLeod, *Inventing the Industrial Revolution: The English Patent System, 1660–1800* (Cambridge: Cambridge University Press,

1988); Liliane Hilaire-Pérez, *L'invention technique au siècle des Lumières* (Paris: Albin Michel, 2000); Zorina B. Khan, *The Democratization of Invention: Patents and Copyrights in American Economic Development, 1790–1920* (Cambridge: Cambridge University Press, 2005); Gabriel Galvez-Behar, *La République des inventeurs: Propriété et organisation de l'innovation en France (1791–1922)* (Rennes: PUR, 2008); and Alain Pottage and Brad Sherman, *Figures of Invention: A History of Modern Patent Law* (Oxford: Oxford University Press, 2010). Since intellectual property remains national law, these studies tend to focus on one country. International histories are more uncommon, but see Fritz Machlup and Edith Penrose, "The Patent Controversy in the Nineteenth Century," *Journal of Economic History* 10, no. 1 (1950): 1–29, and on the early history of the Paris Convention, Yves Plasseraud and François Savignon, *Paris 1883: Genèse du droit unioniste des brevets* (Paris: Litec, 1983). Sam Ricketson and Jane Ginsburg, *International Copyright and Neighbouring Rights: The Berne Convention and Beyond* (New York: Oxford University Press, 2006), is the standard treatise on the history of the Berne Convention. For the particular dilemmas facing scientists, see Paul Lucier, "Court and Controversy: Patenting Science in the Nineteenth Century," *British Journal for the History of Science* 29, no. 2 (1996): 139–54. Stathis Arapostathis and Graeme Gooday show in *Patently Contestable: Electrical Technologies and Inventor Identities on Trial in Britain* (Cambridge, MA: MIT Press, 2013) just how much controversy innovation could entail in court during this period.

Several studies on gender and patenting reveal the extent of the sexing mechanisms of the innovating "he," for instance Deborah J. Merritt, "Hypatia in the Patent Office: Women Inventors and the Law, 1865–1900," *American Journal of Legal History* 35, no. 3 (1991): 235–306; and B. Zorina Khan, "Married Women's Property Laws and Female Commercial Activity: Evidence from United States Patent Records, 1790–1895," *Journal of Economic History* 56, no. 2 (1996): 356–88. Melissa J. Homestead has pursued a similar approach in respect to copyright in her *American Women Authors and Literary Property, 1822–1869* (Cambridge: Cambridge University Press, 2005).

On the *Comptes rendus* and the Académie, see Maurice Crosland, *Science under Control: The French Academy of Sciences, 1795–1914* (Cambridge: Cambridge University Press, 1992); Harry W. Paul, *From Knowledge to Power: The Rise of the Science Empire in France, 1860–1939*

(Cambridge: Cambridge University Press, 1985), and also the contribution by Bruno Latour and Paolo Fabbri, "La rhétorique de la science," *Actes de la recherche en sciences sociales* 13 (1977): 81–95.

On *vulgarisation* and the specialized journal boom in France, see *La science pour tous: Sur la vulgarisation scientifique en France de 1850 à 1914*, ed. Bruno Béguet (Paris: Bibliotheque du conservatoire national des arts et métiers, 1990), and Vincent Duclert and Anne Rasmussen, "Les revues scientifiques et la dynamique de la recherche," in *La belle époque des revues, 1880–1914*, ed. Jacqueline Pluet-Despatin, Michel Leymarie, and Jean-Yves Mollier (Paris: IMEC, 2002), 237–54. Bernadette Bensaude-Vincent deploys a more theoretical approach in "A Historical Perspective on Science and Its 'Others,'" *Isis* 100 (2009): 359–68, and *La science contre l'opinion: Histoire d'un divorce* (Paris: Le Seuil, 2003). In a similar vein, see also Stephen Hilgartner, "The Dominant View of Popularization: Conceptual Problems, Political Uses," *Social Studies of Science* 20, no. 3 (1990): 519–39. Rima D. Apple, Gregory J. Downey, and Stephen L. Vaughn, eds., *Science in Print: Essays on the History of Science and the Culture of Print* (Madison: University of Wisconsin Press, 2012), is a good introduction to the multifarious relationship between science and print.

Richard Yeo, "Alphabetical Lives: Scientific Biography in Historical Dictionaries and Encyclopedias," in *Telling Lives in Science* (139–69), addresses the changing role of the biographical entry in reference works historically. Apart from *Dictionary of Scientific Biography*, ed. Charles Coulston Gillespie (New York: Scribner's, 1980), Marie Curie obviously figures in almost any encyclopedia or multivolume dictionary on twentieth-century science. To get some sense of how these books have helped create the Curie legacy, see Abraham Pais's entry in *Out of the Shadows: Contributions of Twentieth-Century Women to Physics*, ed. Nina Byers and Gary Williams (Cambridge: Cambridge University Press, 2006), 43–55, and Marilyn Bailey Ogilvie's entry on Curie in *The Biographical Dictionary of Women in Science: Pioneering Lives from Ancient Times to the Mid-Twentieth Century*, ed. Marilyn Bailey Ogilvie and Joy Harvey (New York: Routledge, 2000), 311–17.

CHAPTER TWO

First awarded in 1901, the Nobel Prize quickly became famous in its own right. With its current importance undiminished—despite newcomer prizes carrying heftier sums of money—one would expect innumerable books on the cultural, scientific, and economic impact of the Nobel Prize. But the impression is that of scarcity. The year of the centenary, *Minerva* published a special issue (no. 4 [2001]) containing several interesting contributions. Elisabeth Crawford's books remain standard works and provide crucial behind-the-scenes information on, for instance, nominating processes. For Curie's period, see *The Beginnings of the Nobel Institution: The Science Prizes, 1901–1915* (Cambridge: Cambridge University Press, 1984) and also Harriet Zuckerman, "Nobel Laureates in Science: Patterns of Productivity, Collaboration, and Authorship," *American Sociological Review* 32, no. 3 (1967): 391–403. Karin Blanc, *Marie Curie et le Nobel* (Uppsala: Uppsala Studies in History of Science 26, 1999), is the most comprehensive study to date on Curie's 1911 Nobel, providing a number of firsthand sources on the Langevin affair and its repercussions in the Swedish science community. Gustav Källstrand's Ph.D. thesis, *Medaljens framsida: Nobelpriset i pressen 1897–1911* (Stockholm: Carlsson Bokförlag, 2012), focuses on the relationship between the Nobel Prize and the Swedish press and devotes a section (274–83) to the coverage of the Langevin scandal.

For a general introduction to French society during the belle époque see Michel Winock, *La Belle Époque: La France de 1900 à 1914* (Paris: Perrin, 2002). Edward Berenson, Vincent Duclert, and Christophe Prochasson, *The French Republic: History, Values, Debate* (Ithaca, NY: Cornell University Press, 2011), is an valuable encyclopedic resource for anything French, from broader surveys of periods to essential themes of relevance for this chapter, such as family life, anti-Semitism, gender. For a discussion on the gendered lifestyle of scientific couples in general, see Staffan Bergwik, "An Assembled Affinity of Science and Home: The Gendered Lifestyle of Svante Arrhenius and Early Twentieth Century Physical Chemistry," *Isis* 105, no. 2 (2014), 265–91.

The literature on celebrity culture is vast and spans related concepts such as fame, renown, and charisma. Leo Braudy's *The Frenzy of Renown: Fame and Its History* (New York: Vintage Books, 1997) and Richard Schiekel, *Intimate Strangers: The Culture of Celebrity* (New York: Doubleday, 1985), are key texts, with Fred Inglis, *A Short History of Celebrity* (Princeton, NJ: Princeton University Press, 2010) a more concise option. Aaron Jaffe and Jonathan Goldman, eds., *Modernist Star Maps: Celebrity, Modernity, Culture* (Farnham, UK: Ashgate, 2010), contains several illustrative case studies. As I have tried to argue, scientists are perhaps not among those we immediately associate with the term celebrity, but exceptions include the obvious Albert Einstein as well as Charles Darwin; see Janet Browne's "Looking at Darwin: Portraits and the Making of an Icon," *Isis* 100, no. 3 (2009): 542–70, and "Charles Darwin as a Celebrity," *Science in Context* 16, no. 1 (2003): 175–94, the last published in a special issue dealing with persona and celebrity edited by Lorraine Daston and Otto Sibum. To return briefly to the French and Anglo-American styles in respect to biography and science, Nathalie Heinich's "La culture de la célébrité en France et dans les pays Anglophones: Une approche comparative," *Revue française de sociologie* 52, no. 2 (2011): 352–72, is worth mentioning because of its interesting take on the cultural differences vis-à-vis the emergent field of celebrity studies.

Lenard Berlanstein's work has been particularly important for this chapter, for instance "Historicizing and Gendering Celebrity Culture: Famous Women in Nineteenth-Century France," *Journal of Women's History* 16, no. 4 (2004): 65–91, and "Selling Modern Femininity: *Femina*, a Forgotten Feminist Publishing Success in Belle Époque France," *French Historical Studies* 30, no. 4 (2007): 623–49. Apart from his already mentioned *The Trial of Madame Caillaux*, Edward Berenson's co-edited volume with Eva Giloi, *Constructing Charisma: Celebrity, Fame, and Power in Nineteenth-Century Europe* (New York: Berghahn Books, 2010), provided important insights into the gendered nature of French "publicness." Other contributions adding to my understanding of the role of celebrity culture at the time include Mary Louise Roberts, *Disruptive Acts: The New Woman in Fin-de-siècle France* (Chicago: University of Chicago Press, 2002), and for an earlier period Claire Brock, *The Feminization of Fame, 1750–1830* (London: Palgrave Macmillan, 2006). For the special case of Eusapia Palladino, see Christine Blondel, "Eusapia Palladino: La méthode expérimentale et la 'diva des savants,'" in Bernadette Bensaude-Vincent and Christine Blondel, eds., *Des savants face à l'occulte, 1870–1940* (Paris: Éditions la Découverte, 2002). Celebrity worship is not that far from hero worship, and for a very useful discussion of the belle époque cult of the hero, see Venita Datta, *Heroes and Legends of Fin-de-Siècle France: Gender, Politics, and National Identity* (Cambridge: Cambridge University Press, 2011), and Paul Gerbod, "L'éthique héroïque en France," *Revue historique* 268 (1982): 409–29.

On the crucial role of the press and journalism in promoting celebrity culture more generally, see Charles L. Ponce de Leon, *Self-Exposure: Human-Interest Journalism and the Emergence of Celebrity in America, 1890–1940* (Chapel Hill: University of North Carolina Press, 2002). I enjoyed particularly Marcel LaFollette's work on the relationship between science and the press. See primarily his *Making Science Our Own: Public Images of Science, 1910–1955* (Chicago: University of Chicago Press, 1990) and with special emphasis on women: "Eyes on the Stars: Images of Women Scientists in Popular Magazines," *Science, Technology and Human Values* 13, nos. 3–4 (1988): 262–75.

Lawrence Badash, "Radioactivity before the Curies," *American Journal of Physics* 33 (1965): 128–35, and "Radium, Radioactivity, and the Popularity of Scientific Discovery," *Proceedings of the American Philosophical Society* 122, no. 3 (1978): 145–54, are both excellent accounts of the early radium craze. For a comparison with the X-ray craze, see Sylvia Pamboukian, "'Looking Radiant': Science, Photography and the X-Ray Craze of 1896," *Victorian Review* 27, no. 2 (2001): 56–74.

On the controversial relationship between women and the Académie des sciences, see Vesna Crnjanski Petrovich, "Women and the Paris Academy of Sciences," *Eighteenth-Century Studies* 32, no. 3 (1999): 383–90. Londa Schiebinger's work is always essential reading when it comes to gender and science; see *The Mind Has No Sex? Women in the Origins of Modern Science* (Cambridge, MA: Harvard University Press, 1989) and *Nature's Body: Gender in the Making of Modern Science* (New Brunswick, NJ.: Rutgers University Press, 2004). Her "Maria Winkelmann at the Berlin Academy: A Turning Point for Women in Science," *Isis* 78, no. 2 (1987): 174–200, is relevant both for comparing Curie's fate with that of another female applicant at another European academy, and also for understanding the role of widows. For Hertha Ayrton's application to the Royal Society, see Joan Mason, "Hertha Ayrton (1854–1923) and the Admission of Women to the Royal Society of London," *Records of the Royal Society* 45, no. 2 (1991): 201–20, and on her scientific writing and patents, James J. Tattersall and Shawnee L. McMurran, "Hertha Ayrton: A Persistent Experimenter," *Journal of Women's History* 7, no. 2 (1995): 86–112.

For the history of the French press, see Claude Bellanger et al., *Histoire générale de la presse française*, especially vol. 3: *De 1871 à 1940* (Paris: PUF, 1972); Christoph Charle, *Le siècle de la presse, 1830–1939* (Paris: Seuil, 2004); and Christian Delporte, "Presse et culture de masse en

France (1880–1914)," *Revue historique* 299, no. 1 (1998): 93–121. All three consider the question of the 1881 press law, but for a more general introduction to the question of libel, slander, and reputation in Anglo-American jurisprudence, turn to David Rolph, *Reputation, Celebrity and Defamation Law* (London: Ashgate, 2008). Another valuable contribution across media and law is Megan Richardson and Julian Thomas, *Fashioning Intellectual Property: Exhibition, Advertising and the Press, 1789–1918* (Cambridge: Cambridge University Press, 2012).

The Dreyfus affair has generated an insurmountable number of books and articles. In French, see Vincent Duclert, *L'affaire Dreyfus* (Paris: Editions Privat, 2010), and with special focus on the involvement of scientists, his "L'engagement scientifique et l'intellectuel démocratique: Le sens de l'affaire Dreyfus," *Politix* 12, no. 48 (1999): 71–94. Christoph Charle and Eva Telkes, *Les professeurs de la Faculté des sciences de Paris: Dictionnaire biographique, 1901–1939* (Paris: Éditions du CNRS, 1989), not only contains valuable biographical data on some of the Curies' closest collaborators and friends, but also lists their political affiliation. Ruth Harris, *The Man on Devil's Island: Alfred Dreyfus and the Affair That Divided France* (London: Penguin Books, 2011), and Louis Begley, *Why the Dreyfus Affair Matters* (New Haven: Yale University Press, 2009), will take curious readers a long way, and Fredrick Brown, *For the Soul of France: Culture Wars in the Age of Dreyfus* (New York: Alfred A. Knopf, 2010), will provide a general overview of the bitter antagonism between the two camps during the period.

For the history of Action Française, see Eugene Weber, *Action Française: Royalism and Reaction in Twentieth-Century France* (Stanford: Stanford University Press, 1962). For a comprehensive introduction in French, see Jacques Prévotat, *L'Action Française* (Paris: PUF, 2004). On the early history of the movement, see Laurent Joly, "Les débuts de l'Action française (1899–1914) ou l'élaboration d'un nationalisme antisémite," *Revue historique* 639 (2006): 695–718. Kate Cambor's *Gilded Youth: Three Lives in France's Belle Époque* (New York: Farrar, Straus and Giroux, 2009) lets us glimpse a more personal side of Léon Daudet.

The crisis of masculinity figures in several of the studies mentioned in the previous chapter concerning the role of women during the belle époque, but on the critical importance of depopulation, see Davis M. Pomfret, "'A Muse for the Masses': Gender, Age, and Nation in France, Fin de Siècle," *American Historical Review* 109, no. 5 (2004): 1439–74, and

Karen Offen, "Depopulation, Nationalism, and Feminism in Fin-de-Siècle France," *American History Review* 89, no. 3 (1984): 648–76. When it comes to dueling, traditions vary across Europe. For a general history, see V. G. Kiernan, *The Duel in European History: Honor and the Reign of Aristocracy* (Oxford: Oxford University Press, 1989); on Germany, Ute Frevert, *Men of Honour: A Social and Cultural History of the Duel* (London: Polity Press, 1995); on France, Robert A. Nye, *Masculinity and Male Codes of Honor in Modern France* (Oxford: Oxford University Press, 1993), and Jean-Noël Jeanneney, *Le duel: Une passion française, 1789–1914* (Paris: Tempus Perrin, 2011); on Italy, Steven C. Hughes, *Politics of the Sword: Dueling, Honor, and Masculinity in Modern Italy* (Columbus: Ohio State University Press, 2007); on Sweden, Henning Österberg, *Korsade klingor: Om fäktningens och duellerandets historia* (Stockholm: Atlantis, 2010).

CHAPTER THREE

Although Curie's U.S. tours in 1921 and 1929 are described in most biographies, as is her relationship with Meloney, the extensive Curie-Meloney correspondence at the BnF has left surprisingly few traces in scholarly literature. I have searched for but not found a biography of Missy Brown Meloney, who, despite her many years as an important editor, seems to have escaped any sustained consideration. Her story needs to be told, and when it is (she died in 1943 so her writing fell into the "public domain" in 2013), should prove a fascinating read. There is a very brief biographical sketch by Helen Rogers Reid, "Missy Meloney" (31–34), as well as an overview by Edith H. Quimby, "The Marie Curie Correspondence with Marie Mattingly Meloney" (15–23), in the *Columbia Library Columns* 11, no. 2 (1962), a special issue published on the occasion of William Brown Meloney Jr.'s bequest to Columbia University Library of several of Curie's letters to his mother. For an overview of the contents of the Marie Mattingly Meloney papers, 1891–1943, at Columbia University Library, see http://www.columbia.edu/cu/lweb/archival/collections/ldpd_4079096/.

Julie des Jardins, *The Madame Curie Complex: The Hidden History of Women in Science* (New York: Feminist Press, 2010), devotes one chapter to Curie's two U.S. visits, but does not make use of the Curie papers at all. Margaret W. Rossiter, *Women Scientists in America: Struggles and Strategies to 1940* (Baltimore: Johns Hopkins University Press, 1982), is a valuable source for understanding the role of women in U.S. academia at the time but considers Curie only very briefly. On the U.S. radium industry at the time, see Maria Rentetzi, "The U.S. Radium Industry: Industrial In-house Research and the Commercialization of Science," *Minerva* 46, no. 4 (2008): 437–62.

Marcel Mauss, *Essais sur le don: Forme et raison de l'échange dans les sociétés archaïques* (Paris: PUF, 2012), remains a classic work on the gift, but Lewis Hyde, *The Gift: Imagination and the Erotic Life of Property* (New York: Random House, 1983), and especially Marilyn Strathern, *The Gender of the Gift: Problems with Women and Problems with Society in Melanesia* (Berkeley: University of California Press, 1998), focus especially on gender. James Laidlaw discusses the concept of a "free" gift in his "A Free Gift Makes No Friends," in *The Question of the Gift: Essays across Disciplines*, ed. Mark Osteen (London: Routledge, 2002), 45–66.

On Einstein's role in celebrity culture and his relation with the United States, see Alan J. Friedman and Carol C. Donley, *Einstein as Myth and Muse* (Cambridge: Cambridge University Press, 1985), and Marshall Missner, "Why Einstein Became Famous in America," *Social Studies of Science* 15, no. 2 (1985): 267–91. David E. Rowe, "Einstein and Relativity: What Price Fame?" *Science in Context* 25, no. 2 (2012): 197–246, is a more recent and substantial contribution in the same genre. For popular accounts of radium, see David I. Harvie, *Deadly Sunshine: The History and Fatal Legacy of Radium* (The Mill, Gloucestershire: Tempus, 2005), and Jean-Marc Cosset and Renaud Huynh, *La fantastique histoire du radium: Quand élément radioactif devient potion magique* (Rennes: Editions Ouest-France, 2011).

CHAPTER FOUR

On the history of the Académie de médecine, see George Weisz, *The Medical Mandarins: The French Academy of Medicine in the Nineteenth and Early Twentieth Century* (Oxford: Oxford University Press, 1995). Concerning the unpatentability of drugs and the history of medicines in relation to intellectual property, see a special issue of *History of Technology* 24, no. 2 (2008), containing several interesting contributions, especially Maurice Cassier and Christiane Sinding, "'Patenting in the Public Interest': Administration of Insulin Patents by the University of Toronto" (153–71), and the editorial by Jean-Paul Gaudillière, "How Pharmaceuticals Became Patentable: The Production and Appropriation of Drugs in the Twentieth Century" (99–106).

Stanley W. Pycior has documented Curie's work in the CICI and

League of Nations in "'Her Only Infidelity to Scientific Research': Marie Skłodowska Curie and the League of Nations," *Polish Review* 41, no. 4 (1996): 449–67, and her relationship with Einstein in "Marie Skłodowska Curie and Albert Einstein: A Professional and Personal Relationship," *Polish Review* 44, no. 2 (1999): 131–42. Not much has been written on Curie the public intellectual, but for an exception see Michel Pinault, "Marie Curie, une intellectuelle engagée?" *CLIO: Histoire, femmes et sociétés* 24 (2006): 211–29.

Although primarily focused on France, Jean-Jacques Renoliet, *L'UNESCO oubliée: La Société des Nations et la coopération intellectuelle, 1919–1946* (Paris: Publications de la Sorbonne, 1999), remains the most exhaustive treatment to date of the CICI and the IICI. For a slightly broader take on scientists' interaction with the international political/diplomatic community during the interwar years, see Brigitte Schroeder-Gudehus's *Les scientifiques et la paix: La communauté scientifique internationale au cours des années 20* (Montreal: Les Presses de l'Université de Montreal, 1978), and Daniel Kevles, "'Into Hostile Political Camps': The Reorganization of International Science in World War I," *Isis* 62, no. 1 (1971): 47–60.

Mark Mazower, *Governing the World: The History of an Idea* (London: Penguin, 2013), offers an excellent introduction to many of the ideas on internationalism covered in this chapter, including the international management of scientific information. On that topic more directly, see Alex Csiszar, "Seriality and the Search for Order: Scientific Print and Its Problems during the Late Nineteenth Century," *History of Science* 48, nos. 3–4 (2010): 399–434. A recently published focus section of *Isis*, "Ordering the Discipline: Classification in the History of Science" (104, no. 3 [2013]), contains several relevant contributions, with Stephen P. Weldon's "Bibliography Is Social: Organizing Knowledge in the *Isis* Bibliography from Sarton to the Early Twenty-First Century" (540–50) being especially pertinent for this book. However, nobody has done more to make the fascinating world of Paul Otlet and Henri La Fontaine known and available in English than W. Boyd Rayward. See his *The Universe of Information: The Work of Paul Otlet for Documentation and International Organisation* (Moscow: FID, 1975), *International Organisation and Dissemination of Knowledge: Selected Essays of Paul Otlet* (Amsterdam: Elsevier, 1990), as well as his edited volume *European Modernism and the Information Society: Informing the Present, Understanding the Past* (Aldershot, UK: Ashgate 2008). Rayward has also translated a useful introductory volume originally published in French by the Mundaneum as "Mundaneum: Archives of Knowledge" (2010), available from https://www.ideals.illinois.edu/handle/2142/15431.

Comprehensive biographies of Otlet include Alex Wright's recently published *Cataloging the World: Paul Otlet and the Birth of the Information Age* (Oxford: Oxford University Press, 2014) and Françoise Levie, *L'homme qui voulait classer le monde: Paul Otlet et le Mundaneum* (Brussels: Les Impressions Nouvelles, 2007). On Otlet's idol Melvil Dewey, see Wayne A. Wiegand, *Irrepressible Reformer: A Biography of Melvil Dewey* (Chicago: American Library Association, 1996). The anthology *Cent ans de l'Office international de bibliographie, 1895–1995* (Mons: Mundaneum, 1995) and Sylvie Fayet-Scribe's *Histoire de la documentation en France: Culture, science et technologie de l'information 1895–1937* (Paris: CNRS Éditions, 2000) both paint a broad picture of the Belgian and French documentation movement.

Bernadette Bensaude-Vincent, *Langevin 1872–1946: Science et vigilance* (Paris: Belin, 1987), reveals Paul Langevin's long career as a scientist and intellectual, and Dominique Pestre's *Physique et physiciens en France, 1918–1940* (Paris: Editions des archives contemporaines, 1984) gives an overview of the politics of the French scientific community more generally. On Curie's collaboration with other women scientists, see Marlene F. Rayner-Canham and Geoffrey W. Rayner-Canham, *A Devotion to Their Science: Pioneer Women of Radioactivity* (Quebec: McGill-Queen's University Press, 1997), and on the fate of the women working in the watch factories, Claudia Clark, *Radium Girls: Women and Industrial Health Reform, 1910–1935* (Chapel Hill: University of North Carolina Press, 1997).

On the Great War as a watershed moment in patenting, see Christine MacLeod "Reluctant Entrepreneurs: Patents and State Patronage in New Technosciences, circa 1870–1930," *Isis* 103, no. 2 (2012): 328–39. On a somewhat similar note, but from the perspective of France, is Gabriel Galvez-Behar, "Le savant, l'inventeur et le politique: Le rôle du sous-secrétariat d'État aux inventions durant la première guerre mondiale," *Vingtième siècle, Revue d'histoire* 85 (2005): 103–17. Isabella Löhr, "Le droit d'auteur et la première guerre mondiale: Un exemple de coopération transnationale européenne," *Le mouvement social* 244, no. 3 (2013): 67–80, focuses on copyright and developments in the Berne Union at the time.

Scientific property is still an underresearched topic, at least in comparison with the outpouring of books on copyright and media. For a suggestion on how to approach scientific property historically, see Gabriel Galvez-Behar, "The Propertisation of Science: Suggestions for a Historical Investigation," in *Comparativ: Zeitschrift für Globalgeschichte und vergleichende Gesellschaftsforschung* 21, no. 2 (2011): 80–97.

David Philip Miller, "Intellectual Property and Narratives of Discovery/Invention: The League of Nations Draft Convention on 'Scientific Property' and Its Fate," *History of Science* 46, no. 3 (2008): 299–342, carefully details the Anglo-American response to Ruffini's proposal. General overviews of the various stages of the CICI proposal include Stephen Ladas, "The Efforts for International Protection of Scientific Property," *American Journal of International Law* 23, no. 3 (1929): 552–69, and in a somewhat later period, Thomas R. Ilosvay, "Scientific Property," *American Journal of Comparative Law* 2 (1953): 178–97. For a longer treatment of the *droit de suite*, see Henri Desbois, *Le droit d'auteur en France* (Paris: Dalloz, 1973).

On Tho-Radia see Thierry Lefebvre and Cécile Raynal, "De l'Institut Pasteur à Radio Luxembourg: L'histoire étonnante du Tho-Radia," *Revue d'histoire de la pharmacie* 90, no. 335 (2002): 461–80, and on branding and pharmaceutical trademarks in the United States, Jeremy A. Greene, "What's in a Name? Generics and the Persistence of the Pharmaceutical Brand in American Medicine," *Journal of the History of Medicine and Allied Sciences* 66, no. 4 (2010): 468–506, and "The Materiality of the Brand: Form, Function and the Pharmaceutical Trademark," *History and Technology* 29, no. 2 (2013): 210–26.

EPILOGUE

On the depiction of Marie Curie in the *Madame Curie* movie, see Alberto Elena, "Skirts in the Lab: Madame Curie and the Image of the Woman Scientist in the Feature Film," *Public Understanding of Science* 6, no. 3 (1997): 269–78, and T. Hugh Crawford, "Glowing Dishes: Radium, Marie Curie and Hollywood," *Biography* 23, no. 1 (2000): 71–89. On the portrayal of scientists in movies more generally, see Alberto Elena, "Exemplary Lives," *Public Understanding of Science* 2, no. 3 (1993): 205–23.

For a brief introductory sketch of the careers of the Joliot-Curies, see Michel Pinault, "The Joliot-Curies: Science, Politics, Networks," *History and Technology* 13, no. 4 (1997): 307–24. Derek de la Solla Price, *Big Science, Little Science* (New York: Columbia University Press, 1963), is the source of the "little" and "big" I referred to when trying to describe Meloney's contribution to Curie's research.

Many of the texts I used when discussing legal personhood and Marie Curie's status as a married woman in chapter 1 are relevant also here. However, when it comes to understanding the control of our bodies in a digital present, things have become decidedly more complicated. Although much has changed since 1996, Margaret Jane Radin, *Contested Commodities* (Cambridge, MA: Harvard University Press, 1996), remains a valuable introduction to many of the ethical challenges discussed in this chapter. Because the question of legal control of one's own body, property-wise and morally, has become so intense, the literature on the topic has also exploded. Saru M. Matambanadzo, "The Body, Incorporated," *Tulane Law Review* 87, no. 3 (2013): 457–509, is a recent article that looks into the concept of personhood also from the perspective of the corporation as person. On the Henrietta Lacks story as one of unjust enrichment, see Deleso A. Alford, "HeLa Cells and Unjust Enrichment in the Human Body," *Annals of Health Law* 21, no. 1 (2012): 223–36.

The interment at the Panthéon belongs to the kind of collective heritage making that Pierre Nora discusses in his *Les lieux de mémoire* (Paris, 1997). On such processes and science in particular, see Pnina G. Abir-Am, ed., *La mise en mémoire de la science: Pour une ethnographie historique des rites commémoratifs* (Paris: Editions des archives contemporaines, 1998), and for a particular case study, Patricia Fara, "Isaac Newton Lived Here: Sites of Memory and Scientific Heritage," *British Journal for the History of Science* 33, no. 4 (2000): 407–26.

It is impossible to do justice to the enormous number of books that have been published on the commodification of higher education, the relation between secrecy and disclosure, and the tensions between publishing and patenting in the late modern university. I can mention only a few. General overviews include David Greenberg, *Science for Sale* (Chicago: University of Chicago Press, 2007), and H. Rader, ed., *The Commodification of Academic Research* (Pittsburgh, PA: University of Pittsburgh Press, 2010). Corynne McSherry, *Who Owns Academic Work?* (Cambridge, MA: Harvard University Press, 2001), remains the standard work on intellectual property and research, with A. L. Monotti and Sam Ricketson, *Universities and Intellectual Property* (Oxford: Oxford University Press, 2003), providing a more legalistic option.

With particular emphasis on the importance of the Bayh-Dole Act, see Elizabeth Popp Berman, "Why Did Universities Start Patenting? Institution-Building and the Road to the Bayh-Dole Act," *Social Studies of Science* 38, no. 6 (2008): 835–71; Arti K. Rai and Rebecca S. Eisenberg, "Bayh-Dole Reform and the Progress of Biomedicine," *Law and Contemporary Problems* 66, no. 1 (2003): 289–314; Henry Etzkowitz, "Knowledge as Property: The Massachusetts Institute of Technology and the Debate over Academic Patent Policy," *Minerva* 32, no. 4 (1994): 383–421; Charles Weiner, "Patenting and Academic Research: Historical Case Studies," *Science, Technology and Human Values* 12, no. 1 (1987): 50–62; Grischa Metlay, "Reconsidering Renormalization: Stability and Change in Twentieth-Century Views on University Patents," *Social Studies of Science* 36, no. 4 (2006): 565–97; and Jacob H. Rooksby, "Myriad Choices: University Patents under the Sun," *Journal of Law and Education* 42, no. 2 (2013): 313–26. My own interest in trying to capture the publishing/patenting nexus has particularly been sparked by reading Geof Bowker, "What's in a Patent?" in *Shaping Technology/ Building Society: Studies in Sociotechnical Change*, ed. Wiebe E. Bijker and John Law (Cambridge, MA: MIT Press, 1992), 53–74; Amit Prasad, "The (Amorphous) Anatomy of an Invention: The Case of Magnetic Resonance Imagining (MRI)," *Social Studies of Science* 37, no. 4 (2007): 533–60; and especially, perhaps, Greg Myers, "From Discovery to Invention: The Writing and Rewriting of Two Patents," *Social Studies of Science* 25, no. 1 (1995): 57–105.

For a longer treatment of the Institut Curie's challenge to Myriad Genetics and for other cases involving biotech and IP, see Matthew Rimmer, *Intellectual Property and Biotechnology,* and for a fascinating case study in that genre, Sally Smith Hughes, *Genentech: The Beginnings of Biotech* (Chicago: University of Chicago Press, 2011).

Inspiration for the way I think one might conceptualize the brand in this particular context may be found in Celia Lury's writings on the cultural history of the brand; see *Brands: The Logos of the Global Economy* (London: Routledge 2004) and Rosemary Coombe's *The Cultural Life of Intellectual Properties: Authorship, Appropriation, and the Law* (Durham, NC: Duke University Press, 1998). Although legal scholarship has no problem capturing the branded character of someone like Elvis (see, for instance, John Frow, "Elvis' Fame: The Commodity Form and the Form of the Person," *Cardozo Studies in Law and Literature* 7, no. 2

[2002]: 131–71), and while I have always loved Jessica Litman's wonderful description of confusion in "Breakfast with Batman: The Public Interest in the Advertising Age," *Yale Law Journal* 108, no. 7 (1998–99), fully capturing the complexities of how a name such as Dewey or Curie operates in a "brand-like" setting is less common, despite interdisciplinary contributions in a volume such as Lionel Bently, Jennifer Davis, and Jane C. Ginsburg, eds., *Trade Marks and Brands: An Interdisciplinary Critique* (Cambridge: Cambridge University Press, 2008). A more recent suggestion of how this territory could be explored is Sarah Banet-Weiser, *Authentic™: The Politics of Ambivalence in a Brand Culture* (New York: New York University Press, 2012).

Index

Page numbers in italics refer to figures.

Poland: Curie as Polish, 13, 25, 43,
148, 150, 156; Radium Institute in
Warsaw, 103, 126
polonium, 23, 25, 26–27, 28, 48, 72, 154
Pouillet, Eugène, 16, 33
press and media: circulation and
impact of French press, 50; Curie-
Langevin affair, 59–65, 65–67;
Curie's Académie candidacy and
mass press, 53; Curie's refusal of
interviews and autographs, 135–37,
149–50, 160–61; Curie's suffrage
letter, 17; death of Pierre Curie,
37–38; dueling and journalism, 68;
1881 press law, public or private
persons, 61–62; "Le photographe
est sans pitié," *58*; 1903 Nobel
Prize, 42; 1911, Curie's *annus
horribilis*, 6, 47; photographs and
caricatures, *11, 58, 79, 127*; publication of private correspondence, 66;
radium discovery in popular press,
26; relationship with scientific
institutions, 47. *See also* celebrity
culture and image of the scientist
presse de vulgarisation scientifique, 26
private-public divisions: Curie and
1911 Nobel Prize, 72–73; 1881 press
law, 61–62; persona managed
by Curie, 164; and the press, 67,
73–74; publication of private correspondence, 66, 73; upholding
distinctions, 73–74
Prix Gegner, 48
property rights and scientific property: and bibliography as complimentary, 116 (*see also* Commission
internationale de coopération intellectuelle [CICI]); campaign for
droit du savant for scientists, 111,
124, 125–30, 132–33, 141; CICI priority, 7, 120; of Curie as a widow,
39–40; Curie's (women's) lack of,

16–19, 32–37, 76, 121, 153; discovery
and invention distinction, 128–29,
142, 159; extent of personhood
and, 155; as funding argument, 133,
141–42; international treaties and
revisions, 33–34; language and, 158;
moral vs. property rights, 128. *See
also* intellectual property; patents,
trademarks, and copyright
public-private divisions. *See* private-
public divisions
publishing: authorship vs. ownership,
36; Curie's engagement with U.S.,
134–35; of *La radiologie et la guerre*,
111; multiple author conven-
tion, 23, 159–60; and patenting
dichotomy, 12, 35, 157–58; recognition through, 7, 18–19; "without
reserve" regarding radium, 10–12.
See also patents, trademarks, and
copyright; *Pierre Curie* and Curie's "Autobiographical Notes"
Pujo, Maurice, 64, 67
pure/applied science dichotomy,
12–13, 15–16, 29, 73, 76, 82–83, 88;
in bibliography, 119; overview of,
4, 7. *See also* interest and disinterestedness value systems; scientific
practice

Quinn, Susan, 39

radioactivity, Curie coined term, 1, 24
Radione, 138
Radithor patent medicine, 124
radium: and cancer treatment, 88–89,
110; cost of, 6, 82–85, 90, 103; Curie and legacy, heritage, and succession, 153–54; Curie as "mother"
of, 153–54; curie as standard, 49;
deaths from exposure to, 123–24,
144; discovery and isolation of,
1, 14, 23–24, 29–30; 1898 radium